Resolving Messy Policy Problems

Science in Society Series
Series Editor: Steve Rayner
James Martin Institute, University of Oxford
Editorial Board: Gary Kass, Anne Kerr, Melissa Leach,
Angela Liberatore, Stan Metcalfe, Paul Nightingale, Timothy O'Riordan,
Nick Pidgeon, Ortwin Renn, Dan Sarewitz, Andrew Webster,
James Wilsdon, Steve Yearley

Resolving Messy Policy Problems
Handling Conflict in Environmental, Transport, Health and Ageing Policy
Steven Ney

Marginalized Reproduction
Ethnicity, Infertility and Reproductive Technologies
Lorraine Culley, Nicky Hudson and Floor van Rooij

Business Planning for Turbulent Times
New Methods for Applying Scenarios
Edited by Rafael Ramírez, John W. Selsky and Kees van der Heijden

Vaccine Anxieties
Global Science, Child Health and Society
Melissa Leach and James Fairhead

A Web of Prevention
Biological Weapons, Life Sciences and the Governance of Research
Edited by Brian Rappert and Caitriona McLeish

Democratizing Technology
Risk, Responsibility and the Regulation of Chemicals
Anne Chapman

Genomics and Society
Legal, Ethical and Social Dimensions
Edited by George Gaskell and Martin W. Bauer

Nanotechnology
Risk, Ethics and Law
Edited by Geoffrey Hunt and Michael Mehta

Resolving Messy Policy Problems

Handling Conflict in Environmental, Transport, Health and Ageing Policy

Steven Ney

publishing for a sustainable future

London • Sterling, VA

First published by Earthscan in the UK and USA in 2009

ISBN: 978-1-84407-566-9

Typeset by 4word Ltd, Bristol
Cover design by Susanne Harris

For a full list of publications please contact:

Earthscan
Dunstan House
14a St Cross St
London, EC1N 8XA, UK
Tel: +44 (0)20 7841 1930
Fax: +44 (0)20 7242 1474
Email: earthinfo@earthscan.co.uk
Web: **www.earthscan.co.uk**

22883 Quicksilver Drive, Sterling, VA 20166-2012, USA

Earthscan publishes in association with the International Institute for Environment and
Development

A catalogue record for this book is available from the British Library

Ney, Steven.
  Solving messy policy problems: handling conflicts in environmental, transport, health,
and ageing policy / Steven Ney.
      p. cm.
  Includes bibliographical references and index.
  ISBN 978-1-84407-566-9 (hardback)
  1. Policy sciences. 2. Political planning–Europe. I. Title.
  H97.N494 2009
  320.6–dc22
                                   2008046937

At Earthscan we strive to minimize our environmental impacts and carbon
footprint through reducing waste, recycling and offsetting our CO_2 emissions,
including those created through publication of this book. For more details
of our environmental policy, see www.earthscan.co.uk

The book was printed in the UK by
CPI Antony Rowe, Chippenham.
The paper used is FSC-certified and the inks
are vegetable based.

FSC
Mixed Sources
Product group from well-managed
forests and other controlled sources
Cert no. SGS-COC-2953
www.fsc.org
© 1996 Forest Stewardship Council

To my parents

Norbert and Gill

Contents

List of Figures, Tables and Boxes

Figures

Tables

Boxes

Acknowledgements

The single most important thing that has made this book possible has been the inordinate amount of patience that people have had with me during its production.

First and foremost, I would like to thank Rob West, Claire Lamont and Olivia Woodward at Earthscan for all the support and patience they have shown over the past year.

My colleagues at the Singapore Management University are a constant source of inspiration, encouragement and constructive criticism. They deserve a very special mention.

Thanks also to Michael Thompson and Marco Verweij who have provided invaluable friendship, support and solace. Landis MacKellar at IIASA, too, deserves a big thank you for so generously offering time and resources to finish this book.

However, none of this would ever have been possible without the unswerving support and patience from the home team. Thank you, Birgit, Elliot and Ivy.

List of Acronyms and Abbreviations

ACEA	European Automobile Manufacturers' Association
ACF	Advocacy Coalition Framework
ADHD	Attention Deficit Hyperactivity Disorder
ASECAP	Association Européeanne des Concessionnaires d'Autoroute et d'Ouvrages à Péage
ASH	Action on Smoking and Health
ATTAC	Association pour la Taxation des Transactions pour l'Aide aux Citoyens
BAK	Bundesärztekammer/ German Medical Association
BMA	British Medical Association
CBA	cost–benefit analysis
CDU	Christlich Demokratische Union Deutschlands/ Christian Democratic Union of Germany
CER	Community of European Railway and Infrastructure Companies
CLECAT	European Association for Forwarding, Transport, Logistic and Customs Services
CPR	common pool resource
DB	defined-benefit
DBR	Deutsche Bank Research
DGB	Deutscher Gewerkschaftsbund/ The Confederation of German Trade Unions
DRG	Diagnosis Related Groups
EEA	European Environmental Agency
ETSC	European Transport Safety Council
FCCC	Framework Convention on Climate Change
GCC	Global climate change
GHG	greenhouse gas
GSK	GlaxoSmithKline
IAPO	International Alliance of Patients' Organizations
ICC	International Chamber of Commerce
IFPMA	International Federation of Pharmaceutical Manufacturers and Associations
ILO	International Labour Organization

IPCC	Intergovernmental Panel on Climate Change
ISSA	International Social Security Association
JMA	Japanese Medical Association
JPPI	joint public–private initiatives
MART	Manchester Against Road Tolls
MBI	market-based instrument
MCA	multi-criteria analysis
MNC	Multi National Corporation
NDC	Notional Defined Contribution
OECD	Organization for Economic Cooperation and Development
PAYG	Pay-As-You-Go
PHC	People's Health Charter
PHM	People's Health Movement
PNHP	Physicians for a National Health Program
PPP	public–private partnerships
RTD	Research and Technological Development
SARS	severe acute respiratory syndrome
SoVD	Sozialverband Deutschland
SPD	Sozialdemokratische Partei Deutschlands/ Social Democratic Party of Germany
STD	sexually transmitted disease
TETN	Trans-European Networks
TÜV	Technischer Überwachungsverein
UNICEF	United Nations Childrens' Fund
VDR	Verband Deutscher Rentenversicherer
WHO	World Health Organization

1

Introduction

Policy problems no longer seem to get solved. And yet, paradoxically, we seem to be devoting a lot of time and effort to solving these problems. Peek under any 'policy rock' in any country and you are likely to find an anthill of activity. Politicians are striking deals, making speeches, endorsing this and fighting that; civil servants are implementing, monitoring and adjusting policy; experts are producing and refuting expertise at an increasingly rapid rate; pressure groups are mobilizing support for this or against that proposal; journalists put out copy reporting and commenting on all of this; and citizens are writing letters to their MPs, attending town hall meetings, helping out with the local charity or marching in demonstrations. And all this at great expense.

Yet, none of this seems to make much of an impression on today's policy problems. Contemporary policy challenges are tough. They seem to be uncannily adept at developing immunities to any cure policy-makers have so far administered. Indeed, it would sometimes seem as if the more we try to solve these issues, the further away we seem to move from resolution.

Some of these problems are national. Europeans will recognize rising unemployment, creaking welfare states or divisive migration as permanent fixtures on European policy agendas. In the USA, issues such as abortion, education or health care reform create a lot of heated argument but very little in the way of change. In the more affluent parts of Asia, such as Japan or Taiwan, pension reform is a problem that never seems to go away. In Korea, the so-called 'sunshine policy' – describing the policy of detente between South and North Korea – generates conflict with no real promise of resolution.

Some problems affect many countries and regions at once. Global environmental issues, such as biodiversity conservation, air pollution or, not least, climate change produce questions far quicker than policy-makers are able to come up with answers. The same appears to be true for regulation of the global financial system. Some problems seem to challenge policy-makers at local and at global levels at the same time. For example, the global health crisis calls for policy innovation at both very local and international levels.

Some problems, such as international terrorism or emergent health threats, are new. Others, such as poverty, are old. To complicate matters, new policy challenges have changed beyond recognition the contexts for dealing with old favourites such as unemployment, crime or social policy.

Why, then, are contemporary policy issues so resistant to resolution?

Reformstau

Some commentators blame today's policy processes. Instead of enabling swift and decisive policy action, political systems – particularly pluralist democracies – ensnare decision-makers in slow, circular and mind-bogglingly complicated policy processes. Typically, bouts of unsightly bickering between self-serving politicians, officials and lobbyists produce limp compromises falling far short of rational policy solutions. Change, if it takes place at all, progresses at a glacial pace. For example, in Germany, the think-tank Deutsche Bank Research (DBR) argues that:

> ... *rules and checks and balances that are supposed to bring about stability have altered the whole system in a way that it has become increasingly 'change-resistant', exactly at a time when the global environment (globalization, fall of the Iron Curtain) and some of the system's internal parameters (demography) are going through massive structural change. (Bergheim et al, 2003, p9)*

German political commentators have called this Reformstau – a backlog or congestion of urgent reforms necessary to revitalize our societies.

The potential consequences of reform-jam, we are told, are anything but trivial. Repeated disappointments and the perceived inability to deal with urgent policy problems corrode trust in political and policy-making institutions (OECD, 2001; European Commission, 2001). The European Commission (2001) sees in this a dangerous paradox:

> *On the one hand, Europeans want them [policy-makers] to find solutions to the major problems confronting our societies. On the other hand, people increasingly distrust institutions and politics or are simply not interested in them. (p3)*

Decreasing electoral participation and the worrying electoral success of right-wing extremist parties across Europe point to a tangible sense of disappointment with political institutions and democracy.

But other parts of the world are not much different. In the USA, inability to solve old challenges, and fear of new problems – particularly terrorism – have led the US voter to permit massive government infringements of civil liberties. In less affluent parts of the world, in turn, the inability of policy-makers to fulfil the most basic of needs – usually because of rampant corruption – has undermined fledgling democracies across the developed world. In countries such as Thailand or Pakistan political change is often accompanied by violence and civil unrest.

What gets in the way of solving problems, thinkers such as George Tsebelis, Kent Weaver, Paul Pierson and many others contend, is divisive and unnecessary policy conflict. In policy-making, so the argument goes, conflict reflects an underlying imbalance between two incommensurable activities: rational policy-making and pluralist politics. On this view, policy-making is about deploying rational scientific methods to solve objective social problems. Politics, in turn, is about mediating contending opinions, perceptions and world-views. While the former conquers social problems by marshalling the relevant facts, the latter creates democratic legitimacy by negotiating conflicts about values. It is precisely this value-based conflict that distracts from rational policy-making. At best, deliberation and argument slow down policy processes. At worst, pluralist forms of conflict resolution yield politically acceptable compromises rather than rational policy solutions.

Pluralist democracies, more than any other political system, encourage value-driven conflict. Here, so the argument goes, organized interests obtain power over the policy process disproportionate to either their size or actual socio-economic influence (Pierson, 1996, 2001; Bonoli, 2000; Tsebelis, 2002; Bergheim et al, 2003). Strategically placed at critical junctures in the decision-making process (so-called 'veto-points'), these policy actors (so-called 'veto-players') can hold to ransom any policy initiative perceived to jeopardize their interests. In political institutions with many veto-points and veto-players, Tsebelis argues, changes from the status quo are unlikely. What little policy change does take place is typically accompanied by squabbling and haggling as governments buy off obstructive veto-players (Bonoli, 2000; Leibfried and Obinger, 2001).

For these thinkers, unclogging congested policy processes means curtailing the influence of veto-players. In practice, this has meant reducing policy conflict at both the institutional and ideational level (OECD 2001). Box 1.1 provides a brief overview.

Box 1.1 *Controlling veto players*

At the structural level, policy actors across Europe are currently debating how best to streamline and simplify political institutions at regional, national and European level (*Österreichkonvent, Konvent für Deutschland, European Convention*). Here, advocates suggest first separating the appropriate policy-making competences at different levels of governance and, second, paring down substantive policy input across different levels of governance to the merely advisory (the subsidiarity principle) (Strohmeier, 2003). Moreover, advocates also urge policy-makers to shake up ossified power relations within levels of governance. Policy proposals include changes to electoral systems (Strohmeier, 2003), reform of socio-economic decision-making (Tàlos and Kittel, 2001; Bergheim et al, 2003) and realignments in the horizontal separation of power towards the executive (Herzog 2004; European Commission, 2004). In either case, the expressed aim is to reduce the influence of veto-players by curtailing their ability to generate policy conflict.

At the level of ideas, policy actors plan to contain policy conflict with sophisticated knowledge management systems. By basing policy on objective evidence about 'what works', policy actors hope to create 'a common policy focus, [thereby] encouraging participation and mutual understanding ...' among policy actors (Strategic Policy-Making Team, 1999). In this view, values are not only detrimental to understanding and participation, but policy positions based on values are '... likely to fail because they may not be grounded in the economic, institutional and social reality of the problem' (The Urban Institute, 2003, p2). By definition, arguments that challenge the prevalent perception of 'what works' can be safely ignored because they must be based on values rather than evidence. In this way, knowledge management systems designed to bring about 'evidence-based policy-making' help control the policy agenda by narrowing the scope of permitted problems and solutions in the debate.

Messy policy problems

Not everyone agrees with the Reformstau argument. Perhaps, other commentators suggest, it is something about contemporary policy problems that makes them difficult to solve. As the Cabinet Office in the UK points out,

> ... the world for which policy-makers have to develop policies is becoming increasingly complex, uncertain and unpredictable. The electorate is better informed, has rising expectation and is making growing demands for services tailored to their individual needs. Key policy issues, such as social exclusion and reducing crime, overlap and have proved resistant to previous attempts to tackle them, yet the world is increasingly inter-connected and inter-dependent. Issues switch quickly from the domestic to the international arena and an increasingly wide diversity of interests needs to be coordinated and harnessed. Governments across the world need to be able to respond

quickly to events to provide the support that people need to adapt to change and that businesses need to prosper. In parallel with these external pressures, the government is asking policy-makers to focus on solutions that work across existing organizational boundaries and on bringing about change in the real world. Policy-makers must adapt to this new, fast-moving, challenging environment if public policy is to remain credible and effective. (Strategic Policy-Making Team, 1999)

Contemporary policy problems, then, are messy. As early as 1973, Horst Rittel and Melvin Webber (1973) recognized that policy-makers could probably not approach what they called 'wicked problems' with the same toolbox that had served them so well for less unruly issues (see Box 1.2). On the one hand, messy challenges are complex. Contemporary policy problems no longer have clear causes but rather a whole host of loosely connected and interrelated factors. Developments in one seemingly unrelated policy field, say education, may impinge in unpredictable and intricate ways on policy realities of another policy field, say, health policy. For example, the World Health Organization (WHO) (2000) and World Bank (1993) argue that health outcomes in developing countries will depend, in part, on the education of girls. On the other hand, messy policy problems are shrouded in uncertainty. While we have an awful lot of data about issues such as climate change, health care or transport policy, much about the basic variables and factors, let alone how they relate to one another, is simply not known with any degree of certainty.

This book sets out to understand how policy-makers deal with messy or wicked policy problems. It does so by looking closely at the value-driven conflict messy policy problems generate. Somewhat against the grain of received wisdom, the following chapters argue that conflict about messy issues is not a distracting nuisance to rational policy-making. On the contrary, this book suggests that value-driven conflict is not only inevitable but also a crucial resource for dealing with messy policy challenges.

The book demonstrates this by dissecting policy conflict about four different problems: a brief overview of climate change in this chapter followed by more detailed studies of transport (Chapter 3), ageing (Chapter 4) and health (Chapter 5). Although all four policy problems are very different from one another, all of them are messy. In each issue area, policy actors must grapple with challenges the causes of which are either poorly understood or highly contentious or, for the most part, both. Each policy issue depends on the complex interplay of a wide range of factors and variables. Moreover, proposed solutions in each issue area spill-over into other policy domains in unpredictable ways.

The remainder of this introductory chapter provides a brief overview of the book's argument. The following section describes the framework for analysing policy-making about messy problems. To illustrate how the approach works, this chapter applies the framework to the debate about global climate change.

Box 1.2 *Wicked policy problems*

For Rittel and Webber (1973), ten 'distinguishing characteristics' define 'wicked' (as opposed to 'tame') policy problems:

1   'There is no definite formulation of a wicked problem.' Any definition of a wicked problem is inherently uncertain and invariably contested.

2   'Wicked problems have no stopping rule.' Whether or not a wicked problem has been solved is as open to contention as the definition of a wicked problem.

3   'Solutions to wicked problems are not true-or-false, but good-or-bad.' There are no absolute and objective criteria to judge proposed solutions to wicked problems. Thus, assessment of these solutions is a matter of judgement and interpretation.

4   'There is no immediate and no ultimate test of a solution to a wicked problem.' Solutions to wicked problems create 'waves of consequences over an extended – virtually and unbounded – period of time' (Rittel and Webber, 1973, p163). For this reason, the evaluation criteria of any proposed solution will change over time.

5   'Every solution to a wicked problem is a "one-shot operation"; because there is no opportunity to learn by trial and error, every attempt counts significantly.' Wicked problems are persistent, dynamic and adaptable. Not only will solutions change the precise nature of a wicked problem (i.e. the heads it grows back are even more fierce), the solutions themselves will have an impact on the real world and this becomes part of the equation.

6   'Wicked problems do not have an enumerable (or exhaustively describable) set of potential solutions, nor is there a well-described set of permissible operations that may be incorporated into the plan.' Because the causes of wicked problems are essentially uncertain, there are no fixed rules regarding the ingredients that go into the solution of wicked problems.

7   'Every wicked problem is essentially unique.' Regardless of how closely related they may seem, solutions known to work for one wicked problem, say caring for disabled older people, will probably not work for another wicked policy challenge, say caring for disabled young people.

8 'Every wicked problem can be considered to be the symptom of another problem.' Since messy or wicked problems are complex, they are also interrelated in complicated ways.

9 'The existence of a discrepancy representing a wicked problem can be explained in numerous ways. The choice of explanation determines the nature of the problem's resolution.' In other words, the way policy actors define a particular issue shapes the types of policy responses they are likely to prefer. For example, defining secondary smoke as a health and fire hazard has very different policy implications from casting smoking in terms of individual liberties.

10 'The planner has the right to be wrong.' Policy-making, Rittel and Webber argue, is not like science. Policies, unlike scientific hypotheses, have significant direct impacts on people's lives. As a result, the tolerance of mistakes is rather low.

Intractable controversy, policy networks, frames and stories

The Reformstauers are right about one thing. Messy or wicked[1] problems generate persistent and divisive conflict about how best to solve them. The two sociologists Martin Rein and Donald Schön (1993, 1994) have nicely captured the spirit of these conflicts in the term 'intractable policy controversy'. These types of conflict, they continue,

> ... are immune to resolution by appeal to the facts. Disputes of this kind arise around such issues as crime, welfare, abortion, drugs, poverty, mass unemployment, the Third World, the conservation of energy, economic uncertainty, environmental destruction and resource depletion, and the threat of nuclear war. Disputes such as these tend to be intractable, enduring, and seldom finally resolved. (Rein and Schön, 1994, p4)

The danger with intractable policy controversies is that they all too easily degenerate into what the two policy scientists Paul Sabatier and Hank Jenkins-Smith (1993) call a 'dialogue of the deaf'[2] – essentially a shouting match between contending policy actors. It is these dialogues or, more precisely, the failure of dialogue that leads to the policy impasse described as Reformstau.

Knowing how messy policy problems give rise to intractable policy controversies is the key to unravelling the way policy-makers deal with messy issues.

In part, intractable policy controversies are a product of the places in which policy-making occurs today. The past four decades have seen the list of things that we think ought to have some form of public policy grow steadily. On top of classical public duties such as defence, law and order, or public finance, states today manage social welfare, family policy, sports, the arts, health care, education, the youth, the old, women, migrants, the environment, science ... and on and on it goes.

Policy scientists such as R. A. W. Rhodes (1990, 1997) argue that this expansion has also fragmented government. As parts of government specialize on particular policy areas, they need to acquire – through a process of 'buying-in' or 'outsourcing' – the requisite technical knowledge to deal with an issue such as the environment. But then this creates new types of dependencies between governments and non-state actors. Rather than designing rules and ensuring compliance, states today are involved in cooperative relationships with the organizations on which they depend for technical assistance (Ansell, 2000). On the other hand, this process means that today policy-making involves a far wider range of actors and organizations. Instead of being one monolithic institution, the state is scattered across a range of discrete institutional networks that focus on particular policy problems. The collection of these 'policy networks' (Rhodes, 1990, 1997), policy communities (Richardson and Jordan, 1987), issue networks (Heclo, 1978), or policy subsystems (Sabatier and Jenkins-Smith, 1993, 1999) is what Ansell (2000) has called the 'networked polity'. And it is in these pluralist policy networks that policy-makers today have to deal with messy issues.

In part, however, messy issues generate intractable policy controversies because they are complex and uncertain. Complexity and uncertainty do not exactly help effective policy-making. As we shall see presently, issues such as climate change or health care require informed decisions. The popular perception of contemporary policy-making is that hard facts effectively make decisions for policy-makers. Should Europeans ban smoking in public places? The evidence says that secondary smoking is harmful, so the case for a smoking ban is clear.

Or is it? Unfortunately, the evidence claiming that secondary smoking is a health risk is far from conclusive (BBC, 2004). So, what should be an open-and-shut case for banning smoking in public places has been interpreted differently across the European continent. While western Europeans, such as Ireland and the UK, have come down on the side of caution and have banned public smoking, central and southern European countries have taken a more equivocal view. In Germany, the constitutional court revoked a blanket ban in 2008 allowing publicans to permit smoking on their premises under certain conditions. In Austria, a strict ban on smoking in public places never really reached the policy agenda. Policy-makers across Europe have access to the same evidence about secondary smoking, so what is going on?

Messy policy issues are not uncertain and complex because they lack facts. On the contrary, messy policy issues generate reams and terabytes of facts and evidence.[3] The problem is that these facts are not usually all that helpful in determining what to do about messy policy problems. On the one hand, the volume of data on any given messy policy challenge is too large for any policy-maker or group of policy-makers to consider. On the other hand, scientific facts and evidence rarely address the issues that are salient for policy-making. For example, if research were to show that ingesting benzene and benzo(a)pyrene (two components of the particulate matter of cigarette smoke) caused cancers in vital organs of 30 per cent of the mice in a controlled experiment,[4] this would be an interesting finding. In itself, however, it could not justify a ban on smoking in public places.

In order for this bit of evidence to support the case for a smoking ban, someone needs to show how and why it supports the policy initiative. More importantly, it probably needs other, similar bits of evidence to strengthen its impact.

In order to construct this case or argument, the analyst needs to select (what data is relevant) and interpret (how this data is relevant to the practical policy issue in question). What is more, since messy problems are complex and uncertain, the argument needs to be convincing: to be heard, the analyst must resort to rhetoric. All this requires judgement and is a far cry from the idea that 'facts speak for themselves'. For messy policy issues, facts do anything but.

Judgement is guided by shared ideas, values and beliefs. Rein and Schön (1993, 1994) call these shared ideas 'frames'. As the term suggests, these frames help policy actors put evidence and data into a specific context. They provide the criteria for focusing on salient data. They also provide blueprints or templates about how best to interpret the data. In short, frames enable policy actors to make sense of the torrential stream of data, evidence and events that accompany messy policy problems (Rayner, 1991).

In the networked polity, policy scientists have found that individuals tend to work together if they share a particular frame. These groups – sometimes called 'advocacy coalitions' or 'discourse coalitions' – consist of people with a wide range of professional and institutional backgrounds. Their common frame provides them with a shared policy enterprise. For example, we can image the anti-smoking advocacy coalition in Europe to consist, at the core, of anti-smoking activists; these may include such people as the Action on Smoking and Health (ASH) in the UK or the Pro

Rauchfrei eV in Germany. We should not be surprised to find some prominent physicians, perhaps star oncologists from different countries, along with a wide base of GPs across Europe to be part of the coalition. Additionally, the anti-smoking coalition may number some officials at the level of national ministries and the European Commission. Finally, the coalition may also include some politicians, perhaps parliamentarians, and journalists who can carry the argument into the wider public sphere.

Advocacy coalitions try to influence policy in their particular subsystems. In this view, they pursue their common policy enterprise by crafting arguments for or against a particular policy. These arguments, produced by refracting data and evidence through their frame, aim to mobilize other policy actors and galvanize existing support. Advocacy coalitions, through their arguments, tell stories about messy issues. What they are, why they are a problem, who is to blame and what we should do about them. But since advocacy coalitions centre on different frames these stories of messy policy issues are likely to diverge considerably. What is more, these differences cannot be resolved by facts since frames determine what is to count as a fact and what facts are relevant. Neither is this type of conflict amenable to resolution by bargaining: since contending world-views are at issue, there is no basis for negotiation. Policy-making about messy challenges then is an inherently argumentative process in which contending advocacy coalitions pit arguments – plausible and convincing accounts of what is and what should be going on – against each other. This is why conflict about messy issues is inevitably about values and beliefs. And that is also why frame-based conflict about messy issues is inherently intractable.

The following chapters analyse frame-based conflict by reconstructing and comparing contending arguments about messy issues. A useful way of doing just that is to think of the arguments as narratives or stories. These stories are based on specific assumptions (the setting), define the problem (the villains) and propose solutions (the heroes).

The approach compares these policy stories using the typology of frames inspired by the late Dame Mary Douglas (1970, 1982, 1996) and developed by Mike Thompson (Thompson et al, 1990; Thompson et al, 1999). This approach, sometimes called Cultural Theory,[5] is based on the simple insight that the way we live shapes the way we think about the world. The social relations we enter into provide us with the normative and cognitive tools to make sense of messy policy issues. These tools also furnish us with norms to guide the interaction with others. In this view, social structures generate specific frames and interpretive templates. These frames, in turn, help legitimize the preferred social form.

Cultural theorists identify five basic sets of social relations: hierarchy, the market, nested bounded groups, isolation and the hermit. Each of these social relations gives rise to a particular frame: hierarchism, individualism, egalitarianism, fatalism and autonomy. Each of these frames enables its members to select and focus on the salient aspects of messy issues. Here, salience is what helps reproduce and legitimize the preferred form of social relations.

We can use this typology to compare contending advocacy coalitions and the stories they tell. Some coalitions are likely to impose clear and differentiated roles on their members. Members of these coalitions will value order, harmony and process. In other coalitions, members will freely negotiate their relations with one another. These coalitions will emphasize individual liberties, competition and the primacy of the bottom-line. Other coalitions will be well-defined groups that shun internal distinctions; members of these coalitions will stress equality, holism and the ever-present need to speak out against injustice. Members of the last two forms of social relations do not take part in policy debates. Fatalists, isolated as they are, see no reason to participate in politics since whatever they do never seems to amount to much. Hermits, in turn, go out of their way to avoid any social interaction.

The typology of frames, then, is the scalpel for dissecting contending policy stories. By identifying and reconstructing contending policy stories, the typology gauges the distance between contending advocacy coalitions. This is the 'scope of policy conflict'. But since each advocacy coalition defines itself in contradistinction to the other, the contending coalitions are constantly probing, examining and exploring each other's stories. This leads to complicated patterns of agreement and disagreement across advocacy coalition boundaries. This is what we will call the 'structure of policy conflict'. Since each frame is a partial and selective account of any particular messy issue, each policy story has strengths and weaknesses. These, then, are the 'potential impacts' of policy stories.

How does this approach work in practice?

Global climate change

Global climate change (GCC) is the type of messy challenge that seems to have policy-makers stumped. Despite over a decade of negotiation at intergovernmental level, commentators agree that the key policy output to date – the Kyoto Protocol – is disappointing (Verweij et al, 2006; Verweij, 2006). Years of tough bargaining have produced a somewhat timid commitment from governments in industrial countries to cut greenhouse gas (GHG) emissions by an unspectacular 5 per cent. Even if the US government, representing the country with the highest carbon dioxide emissions, had not unilaterally withdrawn from the protocol in 2001, the reduction targets were far below what the Intergovernmental Panel on Climate Change (IPCC) considers necessary for stabilizing atmospheric carbon dioxide concentrations (IPCC, 2001). At this rate, it would require another 30 Kyoto Protocols to even get close (Verweij et al, 2006).

Yet, even within its own rather modest terms, the Kyoto Protocol stands little chance of succeeding. Of the countries that have ratified the Kyoto Protocol, few are actually implementing measures to cut emissions. Those few are unlikely to fulfil their GHG reduction commitments. The European Environmental Agency (EEA) projects that existing climate change policies in EU member states will only be able to cut a projected 1 per cent of their GHG emissions (Gardiner et al, 2003). While additional policy measures may bring the reductions up to 7.7 per cent (Gardiner et al, 2003), it is far from certain that EU member states will implement these carbon dioxide reductions. Given the difficulties to get the Kyoto Protocol up and hobbling, insiders are less than optimistic about the future of the UN's Framework Convention on Climate Change (FCCC) policy process (Verweij et al, 2006).

Critical observers like to think of climate change policy-making as the textbook case of too many veto-players spoiling the rational policy broth (cf. Sprinz and Luterbach, 2001; Rowlands, 2001). Led by US negotiators under the sway of business interests at home, so the argument goes, a coalition of dissenting countries (including China, Australia and, up until recently, Russia) has consistently blocked any attempt at imposing necessary GHG emission reduction targets (Sprinz and Luterbach, 2001). Similarly, developing countries, represented by the G77, have deftly used the UN framework to link global poverty to climate change thereby justifying their exemption from carbon dioxide emission reduction (Bodansky, 2001). In either case, commentators imply, veto-players driven by short-term self interests have used negotiation and compromise to frustrate long-term rational policy based on scientific evidence. Climate change policy, it seems, would be in a much better shape without the distractions of pluralist politics.

But would it be? Using the approach sketched above to map the scope and structure of policy conflict suggests a somewhat different interpretation.

Scope of policy conflict

An analysis of the global climate change policy debate reveals three contending policy stories about global warming (Thompson and Rayner, 1998a, 1998b; Thompson et al, 1999). Each policy story below creates a setting, points to villains and declares its heroes. Depending on the social relations of the advocacy coalition, each story emphasized different aspects of the climate change issue. What is more, each story defines itself in contradistinction to the other policy stories.

Profligacy: an egalitarian tale

The first story begins by pointing to the profligate consumption and production patterns of the North as the fundamental cause of climate change. Rich industrialized countries, so the argument goes, are recklessly pillaging the world's resources with little regard to the well-being of either the planet or the peoples of the poorer regions of the world. Global climate change is more than a problem amenable to quick technical fixes: it is a fundamentally moral and ethical issue.

The setting for this story is a world in which everything is intricately connected with everything else. Whether this concerns human society or the natural world, this story urges us to think of Planet Earth as a single living entity. Environmental degradation then is also an attack on human well-being. Humans, so the argument goes, have successfully deluded themselves that they can live apart from the natural environment. In reality, however, there is no place for humans outside nature and thus no particular reason for considering humans as superior to nature. In short, this story is set in an eco-centric world.

The villain in the profligacy story is the fundamentally inequitable structures of advanced industrial society. In particular, the profit motive and the obsession with economic growth, the driving forces of global capitalism, have not only driven us to the brink of ecological disaster, but have also distorted our understanding of both the natural and the social world. Global commerce and the advertising industry lead us to desire environmentally unsustainable products (bottled water, fast cars and high calorie foods, for example) while our real human needs (living in harmony with nature and with each other) go unfulfilled. What is more, advanced capitalism distributes the spoils of global commerce highly inequitably. This is true within countries (the increasing gap between the rich classes and the poor classes) and among countries (the increasing gap between the affluent countries of the North and the destitute countries of the South). In short, prevailing structural inequalities have led to increasingly unsustainable patterns of consumption and production.

Since everything is connected to everything else, this story maintains, we cannot properly understand environmental degradation unless we see it as a symptom of this wider social malaise. The way humans pollute, degrade and destroy the natural world is merely a very visible indicator for the way they treat each other and particularly the weak members of society. The logic that allows us to fell thousands of square-kilometres of rainforest, to dump toxins in waterways or pollute the air is precisely the same logic that produces racism, misogynism and xenophobia. Tackling one problem inevitably implies tackling all the others.

The heroes of the profligacy story are those organizations and individuals who have managed to see through the chimera of progress in advanced industrial society. They are those groups and persons that understand that the fate of humans is inextricably linked to the fate of Planet Earth. The heroes understand that, in order to halt environmental degradation, we have to address the fundamental global inequities. In short, the heroes of the profligacy policy argument are those organizations of protest such as, most prominently, Greenpeace or Friends of the Earth.

The heroes of this story point to a number of solutions. In terms of immediate policy, the tale of profligacy urges us to adopt the precautionary principle in all cases: unless policy actors can prove that a particular activity is innocuous to the environment, they should refrain from it. The underlying idea here is that the environment is precariously balanced on the precipice of disaster. The story further calls for drastic cuts in carbon dioxide emissions; since the industrialized North produces most carbon dioxide emissions, the onus is on advanced capitalist states to take action.

Yet none of these measures, the story continues, is likely to be fruitful on its own. In order to really tackle the problem of global climate change, we in the affluent North will have to fundamentally reform our political institutions and our unsustainable life-styles. Rather than professional democracies and huge centralized administrations, the advocates of the profligacy story suggest that we decentralize decision-making down to the grass-roots level. Rather than continuing to produce ever-increasing amounts of waste, we should aim at conserving the fragile natural resources we have: we should, in a word, move from the idea of a waste society to the concept of a 'conserve society'. Only then can societies equitably meet real human needs. What are real human needs? Simple, they are the needs of Planet Earth. And this is why Greenpeace seeks:

> ... a world in which the manifest benefits of energy services, such as light, heat, power and transport are equitably available for all: north and south, rich and poor. Only in this way can we create true energy security, as well as the conditions for true human security. (Greenpeace, 2007, p2)

Prices: an individualist story

The second story locates the causes for global climate change in the relative prices of natural resources. Historically, prices have poorly reflected the underlying economic scarcities; the result, plain for all to see, is a relative over-consumption of natural resources.

The setting of the prices tale is the world of markets and economic growth. Unlike the profligacy story, the price diagnosis sees no reason to muddy the conceptual waters with extraneous considerations of social equality. While this is important, the story of prices and property rights understands global climate change as an issue amenable to precise analytical treatment: it is, in short, a technical issue to which we can apply a technical discourse.

Economic growth, far from being a problem, is the sole source of salvation from environmental degradation. Environmental protection, the proponents of this policy argument contend, is a very costly business. In order, then, to be able to foot the huge bill of adjusting to a more sustainable economy, societies will have to command sufficient funds. These funds, in turn, will not materialize from thin air: only economic growth can provide the necessary resources to tackle the expensive task of greening the economy.

The prices tale takes place in a world determined by the invisible hand. Here, individuals know and can precisely rank their preferences. In the world of the prices story, individual pursuit of rational self-interest (economic utility) leads to optimal allocation of resources. If market forces are allowed to operate as they should, resource prices will accurately reflect underlying scarcities: the price mechanism keeps consumption in check. If, of course, someone (usually the misguided policy-maker) meddles with market forces, prices cannot reflect real scarcities: this gives rise to incentives for rational economic actors to over- or under-consume a particular resource.

The villain in the prices story is misguided economic policy. Barriers to international trade, subsidies to inefficient national industries, as well as price and wage floors introduce distortions to the self-regulatory powers of the market. These distortions have historically led markets to place a monetary value on natural resources that belies the true market value. The result, the protagonists of this policy argument maintain, has been wholesale over-consumption and degradation of the natural world.

The heroes of the prices story are those institutions that understand the economics of resource consumption. In the global climate change debate, these institutions comprise players such as the World Bank or the Organization for Economic Cooperation and Development (OECD). The solution to the climate change issue is as simple as its prognosis: in order to successfully face the challenge of global climate change, we have to 'get the prices right'. Unlike the profligacy story, the prices tale sees no necessity to restructure existing institutions. If it is the distortions of global, national and regional market mechanisms that undervalue natural resources, then any climate change policy that fails to remove these distortions is 'fundamentally flawed'. Policy responses must work 'with the market'. Here, concrete policy proposals consist of both general measures, such as the liberalization of global trade, as well as more specific measures, such as carbon taxes or tradable emission permits. For proponents of the prices story, such as the International Chamber of Commerce (ICC):

> ... *the basic challenge is to meet the world's growing demand for energy that is essential to sustain economic growth and improve living standards, while also addressing long-term concerns about greenhouse gas emissions and the environment. (ICC, 2004)*

Proportion: a hierarchical tale

The third policy argument tells a story of uncontrolled and disproportionate development in the world, particularly poorer regions. Rapidly increasing population in the South, this story argues, is placing local and global ecosystems under pressures that are fast becoming dangerously uncontrollable: more people means more resource consumption which inevitably leads to environmental degradation. In the developed North, unregulated markets have led to environmentally imbalanced societies.

The setting of the proportion story differs slightly, but significantly, from the settings in the other two diagnoses. Like the protagonists of the profligacy story, the proportion policy argument maintains that global climate change is a moral issue. Human beings, owing to their singular position in the natural world, are the custodians of Planet Earth; since civilization and technological progress have allowed us to understand the natural world more than other species, we have a moral obligation to apply this knowledge wisely. Unlike the profligacy story, the proportion tale assumes that humans have a special status outside natural processes. Similarly, the proportion story, like the proponents of the pricing argument, contends that human actions are rational. However, unlike the market argument, the proportion story tells us that the sum of individual rational actions can lead to irrational and detrimental outcomes. The proportion story, then, is set in a world that needs rational management in order to prevent society slipping into disorder and decline. Yet, whereas the motive of rational management is an ethical duty to preserve the planet, the means of management are technical.

Economic growth, and the socio-economic system that underpins growth, are necessary components in any global climate change policy response. However, economic growth in itself is no solution: it must be tempered, directed and balanced by the careful and expert application of knowledge and judgement.

The villain in the population tale is uncontrolled and imbalanced development around the globe. In the developing world, uncontrolled population growth threatens the global climate. Since each individual has a fixed set of basic human needs (such as food, shelter, security, etc.) and these needs are fixed at every level of socio-economic development, population increase, other things being equal, must lead to an increase in human needs. Humans, so the story continues, satisfy their basic human needs by consuming resources. It follows that population growth must lead to an increase in resource consumption: more people will produce more carbon dioxide to satisfy their basic needs. Given the limited nature of most resources, population growth must invariably lead to over-consumption and degradation of natural resources.

In the developed North, unregulated markets have created societies that recklessly squander natural resources. Whether it is energy use, transport systems or consumption habits, the people in the affluent North, so the argument goes, have lost their sense of proportion and responsibility – both to others on this planet and to future generations. It is, in short, time to rein in and control this lop-sided development.

The heroes of the proportion story are those institutions with both the organizational capacities (that is, the technical knowledge) and the 'right' sense of moral responsibility for global climate change issues. In short, the global climate change issue should be left to 'experts' situated in large-scale, well-organized administrations. It is only in these institutions that we find the capacity to orchestrate the many components needed to balance development and growth on the planet. In the South, proponents of the proportion story suggest the rational control of population growth. In particular, this means the introduction of family planning and education in the countries most likely to suffer from rapid population growth. In the North, the champions of the proportion story aim to harness, coordinate and integrate the many resources needed for implementing effective climate change mitigation strategies. This is why ...

> ... *the Union is committed to the principle of 'sustainable develop-ment': finding that elusive balance between protecting the environment, ensuring economic progress and social development. Its overall aim is to improve the quality of life and at the same time to protect the environment so that future generations, in all parts of the world, can develop and prosper. (European Commission, 2002, p4)*

Structure of policy conflict about global climate change

The three policy stories about global climate change create complicated patterns of agreement and disagreement between the contending advocacy coalitions. All three advocacy coalitions can agree that global climate change constitutes a challenge that requires some form of response. However, agreement beyond this general proposition disaggregates into pair-wise alliances. Here, members of advocacy coalitions agree on some fundamental principles and general policy measures. At a more concrete level, however, agreement collapses into an intractable controversy as each advocacy coalition fleshes out general policy principles.

Advocates of both the profligacy and prices story agree on the superiority of decentralized responses to global climate change. Proponents of both policy stories champion small-scale and flexible renewable energy sources, local adaptation and mitigation, as well as the relocation of autonomy and choice to the most appropriate level. Similarly, members of both advocacy coalitions are sceptical of policy responses that centralize and concentrate the responsibility for global climate change policy. The overly bureaucratic treaties of the FCCC, as well as unwieldy mechanisms such as the Joint Implementation Mechanism, members of both advocacy coalitions agree, will do little to curb carbon dioxide emissions. Moreover, both advocacy coalitions reject massive investment in large, centralized technologies such as nuclear fusion. Therefore, proponents of the profligacy and prices stories both support policy that encourages and promotes development of renewable forms of energy provision.

Proponents of the population and prices stories, in turn, understand the significance and necessity of economic growth. As we have seen, both advocacy coalitions are in no doubt that curbing global climate change will require resources that only continuing and increased economic growth can deliver. For this reason, members from both advocacy coalitions reject policy proposals that aim to sacrifice economic growth for stringent environmental standards: both proponents of the prices and population stories are critical of applying the strong version of the precautionary principle. Consequently, both advocacy coalitions support policies and measures that can lead to a 'greening' of the world economy.

Advocates of the population and profligacy stories both agree in principle that curbing global climate change is also a matter of taking social responsibility. This includes the international reallocation of resources from rich to poor as well as community-driven environmental protection. Policy actors from these advocacy coalitions pour scorn on policies and measures that absolve the community of any responsibility for global climate change mitigation: voluntary codes and targets by industry, both contend, are not only ineffective but also morally corrosive. For this reason, proponents from both advocacy coalitions favour environmental standards that impose restrictions on unfettered emission of carbon dioxide.

However, this agreement dissolves into intractable disagreement as each advocacy coalition brings incompatible frames to bear on the principles and policy measures. While advocates of the profligacy and prices stories are united in their opposition to centralized policy responses to global climate change, they disagree on the purpose of decentralized solutions. Advocacy coalitions committed to reducing global inequities see renewable energy technologies as a way to decouple energy production and consumption from the capitalist mode of production. Market-oriented coalitions, in turn, see renewable energy sources and flexible technologies as a means of boosting and strengthening economic growth. Similarly, the advocates of the prices and population stories may agree on the need for economic growth. They, however, disagree on how to bring this growth about: whereas hierarchical actors favour the steering of economic growth in an environmentally sound direction, individualists argue for setting up real incentives for individuals and firms to 'go green'. Last, while the proponents of the population and profligacy stories see global climate change mitigation as a moral issue, they bitterly disagree on what type of morality is involved. The egalitarian actors see restoring social responsibility as an act of total socio-cultural renewal in which existing inequities are levelled. Conversely, the hierarchical advocacy coalition understands social responsibility as a stratified system of burden-sharing and problem-solving.

Irreducible plurality: the potential impacts of policy stories

The three stories tell three plausible but conflicting tales about climate change. All three tales use reason and logic to argue their points. None of the tales is 'wrong' in the sense of being implausible or incredible. Yet, at the same time, none of the stories is completely 'right': each argument focuses on those aspects of climate change for which there is a suitable solution cast within the terms of a particular form of organization.

Most importantly, these three policy discourses are not reducible to one another. No one of the policy arguments is a close substitute for either of the others, nor are the stories' proponents likely to agree on the fundamental causes and solutions of global climate change. These policy stories simply cannot be collapsed into one another; rather, thanks to their mutual incompatibility, they frame and define the global climate change issue. Since they implicitly transport a normative argument, namely that of the good life (either in markets, in egalitarian bound-groups, or in hierarchies), they are curiously immune to 'enlightenment' by 'scientific' facts – we cannot, in any scientific sense, 'prove' or 'falsify' policy stories. More pointedly, these stories define what sort of evidence counts as a legitimate fact and what type of knowledge is credible. The profligacy story discounts economic theory as the obfuscation of social inequalities and dismisses rational management as the reification of social relations. The tale of prices views holistic eco-centrism as amateur pop-science and pours scorn on the naive belief in benign control. Last, the population story rejects laissez-faire economic theory as dangerously unrealistic and questions the scientific foundations of more holistic approaches.

This leaves us with a dynamic, plural and argumentative system of policy definition and policy framing that policy-makers can ignore only at their cost. We have seen that each policy story thematizes a pertinent aspect of the climate change debate: very few, apart from extreme hard-liners, would argue that Northern consumption habits, distorted prices or population growth have no impact on global climate change at all. However, as we have seen, each story places a different emphasis on each aspect. Any global climate change policy, then, based only on one or two of these stories, will merely provide a response to a specific aspect of the global climate change problem. It will, in short, provide a partially effective response to the climate change problem.

Left to their own devices, each partial policy solution is vulnerable to its in-built weaknesses. In its focus in global inequities, members of the profligacy story downplay the important question of how to curb carbon dioxide emissions in developing countries. And yet, GHG emissions in countries such as India or China are set to grow by double-digit figures in the coming decades. Proponents of the prices story are so confident in the self-regulating powers of markets that it blinds them to the environmental degradation caused by unregulated resource markets. What is more, proponents of the prices story gloss over the fact that the fruits of economic growth accrue to people with few incentives to protect the environment. Last, champions of the population story place their trust in the competence and moral rectitude of hierarchical institutions and their experts. Yet this trust also makes the institutions so favoured by proponents of the population story vulnerable to bureaucratic inertia and corruption.

Excluding veto-players, then, would expose the climate change regime to surprise and policy failure. It would seem as if all three stories contribute to effective policy in both a positive sense (by providing ideas and solutions) as well as a negative sense (by pointing to the flaws and problems of solutions from other advocacy coalitions).

Mess and plurality

This begs at least two sets of questions.

First, how can this type of analysis help us understand other complex and uncertain issues? We all know that climate change is a uniquely difficult policy challenge. And that is why, some could argue, it needs unique solutions. Solutions, moreover, that have little relevance to policy challenges as diverse as transport, ageing or health. In other words, how does this type of narrative analysis work for other types of policy problems?

Second, what does this irreducible plurality mean for contemporary policy processes? If messy policy problems give rise to conflicting and partially accurate accounts, how do we structure policy-making to respond to these challenges?

These are precisely the questions that the following chapters will address. Before delving into frame-based conflict though, the next chapter discusses the conceptual framework in a little more detail.

Notes

1 The remainder of the book will use the term 'messy' problems rather than Rittel and Webber's term 'wicked'. While the policy issues that the book deals with are certainly messy, I am not sure they are as malicious – as the term 'wicked' implies.
2 This is, of course, a horribly inaccurate and unkind description of the way in which hearing-impaired people communicate. Feminists may argue, that 'lads down the pub' may be a more accurate representation of policy conflict in which everyone holds strong opinions, likes to articulate them at high volume and cares little for what anyone else has to say.
3 Readers in doubt are invited to flick through any of the Intergovernmental Panel on Climate Change's Assessment Reports.
4 This is a fictitious example.
5 Alternative names for this family of approaches include grid/group analysis, neo-Durkheimian theory or theory of plural rationalities.

References

Ansell, C. (2000) 'The Networked Polity: Regional Development in Western Europe', *Governance: An International Journal of Policy and Administration*, vol 13, no 3, pp303–333

Bergheim, S., Neuhaus, M. and Schneider, S. (2003) 'Reformstau – Causes and Remedies', *Current Issues*, no 13, pp1–10

Bodansky, D. (2001) 'The history of the global climate change regime', in D. F. Sprinz and U. Luterbach (eds) *International Relations and Global Climate Change*, MIT Press, Cambridge, MA, pp23–40

Bonoli, G. (2000) *The Politics of Pension Reform: Institutions and Policy Change in Western Europe*, Cambridge University Press, Cambridge

British Broadcasting Corporation (2004) 'Q&A Passive Smoking', http://news.bbc.co.uk/2/hi/health/medical_notes/3235820.stm, accessed 20 October 2008

Douglas, M. (1970) *Natural Symbols: Explorations in Cosmologies*, The Cresset Press, London

Douglas, M. (ed) (1982) *Essays in the Sociology of Perception*, Routledge and Kegan Paul, London

Douglas, M. (1996) *Thought Styles: Critical Essays on Good Taste*, Sage Publications, London

European Commission (2001) 'European Governance: A White Paper', COM(2001) 428 final, Brussels

European Commission (2002) *Choices for a Greener Future: The European Union and the Environment*, Office for Official Publications of the European Communities, Brussels

European Commission (2004) *Summary of the Agreement on the Constitutional Treaty*, Commission of the European Community, Brussels

Gardiner, A., Taylor, P., Carnes, M. and Handley, C. (2003) 'Greenhouse Gas Emission Projections for Europe', *European Environmental Agency Technical Report*, no 77, Copenhagen

Greenpeace (2007) 'Renewable energy and climate change', www.greenpeace.org/raw/content/international/press/reports/renewable-energy-and-climate.pdf, accessed 22 October 2008

Heclo, H. (1978) 'Issue Networks and the Executive Establishment', in A. King (ed) *The New American Political System*, American Enterprise Institute, pp87–124

Herzog, R. (2004) 'Ein neuer Anlauf zur Föderalismusreform ist nötig', *Konvent für Deutschland*, Berlin

ICC (2004) 'Business Perspectives on a Long-Term International Policy Approach to Address Global Climate Change' www.iccwbo.org/collection4/folder165/id581/printpage.html?newsxsl=&articlexsl=, accessed 13 April 2008

Leibfried S. and Obinger, H. (2001) 'Welfare State Futures: An Introduction' in S. Leibfried and H. Obinger (eds) *Welfare State Futures*, Cambridge University Press, Cambridge, pp1–8

OECD (2001) 'Government of the Future', *PUMA Policy Brief*, no 9, Paris

Pierson, P. (1996) 'The New Politics of the Welfare State', *World Politics*, vol 48, pp143–179

Pierson, P. (ed) (2001) *The New Politics of the Welfare State*, Oxford University Press, Oxford

Rayner, S. (1991) 'A Cultural Perspective on the Structure and Implementation of Global Environmental Agreements', *Evaluation Review*, vol 15, no 1, pp75-102

Rhodes, R. A. W. (1990) 'Policy Networks: A British Perspective', *Journal of Theoretical Politics*, vol 2, no 3, pp293–317

Rhodes, R. A. W. (1997) *Understanding Governance: Policy Networks, Governance, Reflexivity and Accountability*, Open University Press, Buckingham

Richardson, J. J. and Jordan, A. G. (1987) *British Politics and the Policy Process: An Arena Approach*, Allen & Unwin, London

Rittel, H. and Webber, M. (1973) 'Dilemmas in a General Theory of Planning', *Policy Sciences*, vol 4, pp155–169

Rowlands, I. (2001) 'Classical Theories of International Relations', in D. F. Sprinz and U. Luterbach (eds) *International Relations and Global Climate Change*, MIT Press, Cambridge, MA, pp34–66

Sabatier, P. and Jenkins-Smith, H. (1993) *Policy Change and Learning: An Advocacy Coalition Approach*, Westview Press, Boulder, CO

Sabatier, P. and Jenkins-Smith, H. (eds) (1999) *Theories of the Policy Process*, Westview Press, Boulder, CO

Schön, D. and Rein, M. (1993) 'Reframing Policy Discourse', in F. Fischer and J. Forester (eds) *The Argumentative Turn in Policy Analysis and Planning*, Duke University Press, Durham, NC

Schön, D. and Rein, M. (1994) *Frame Reflection: Towards the Resolution of Intractable Policy Controversies*, Basic Books, New York, NY

Sprinz, D. F. and Luterbach, U. (ed) (2001) *International Relations and Global Climate Change*, MIT Press, Cambridge, MA

Strategic Policy-Making Team (1999) *Professional Policy-Making for the Twenty-First Century*, The Cabinet Office, London

Strohmeier, G. A. (2003) 'Zwischen Gewaltenteilung und Reformstau: Wie viele Vetospieler braucht das Land', *Aus Politik und Zeitgeschichte*, vol 51, 15 Dec, pp17–22

Tàlos, E and Kittel, B. (2001) *Gesetzgebung in Österreich: Netzwerke, Akteure und Interaktionen in politischen Entscheidungsprozessen*, WUW-Universitäts-Verlag, Vienna

Thompson, M. and Rayner, S. (1998a) 'Cultural Discourses' in S. Rayner and E. Malone (eds) *Human Choice and Climate Change, Volume 1: The Societal Framework*, Batelle Press, Columbus, OH

Thompson, M. and Rayner, S. (1998b) 'Risk and Governance Part I', *Government and Opposition*, vol 33, no 2, pp139–166

Thompson, M., Wildavsky, A. and Ellis, R. (1990) *Cultural Theory*, Westview Press, Boulder, CO

Thompson, M., Rayner, S. and Ney, S. (1998) 'Risk and Governance Part II', *Government and Opposition*, vol 33, no 3, pp330–354

Thompson, M., Grenstad, G. and Selle, P. (1999) *Cultural Theory as Political Science*, Routledge, London

Tsebelis, G. (2002) *Veto Players: How Political Institutions Work*, Princeton University Press, Princeton, NJ

Urban Institute (2003) *Beyond Ideology, Politics and Guesswork: the Case for Evidence-Based Policy*, The Urban Institute, Washington DC

Verweij, M. (2006) 'Is the Kyoto Protocol Merely Irrelevant, or Positively Harmful, for the Efforts to Curb Climate Change' in M. Verweij, and M. Thompson (eds) *Clumsy Solutions for a Complex World*, Palgrave, Basingstoke

Verweij, M., Douglas, M., Ellis, R., Engel, C., Hendriks, F., Lohmann, S., Ney, S. and Rayner, S. (2006) 'Clumsy Solutions for a Complex World: The Case of Climate Change', *Public Administration*, vol 84, no 4, pp817–843

WHO (2000), *The World Health Report 2000: Health Systems – Improving Performance*, WHO, Geneva

World Bank (1993) *World Development Report 1993: Investing in Health*, World Bank, Washington DC

2

Understanding Policy Conflict

Introduction

What makes problems such as the climate change or demographic ageing so messy is the tenacity of the public conflicts they generate. Whether it is climate change or pension reform, all messy issues create persistent and divisive conflict about how best to solve them. As we saw in the introductory chapter, the two social scientists Donald Schön and Martin Rein (1993, 1994) call these types of conflicts 'intractable policy controversies'. Immune to either enlightenment or negotiation, 'intractable policy controversies' are adept at finding new sources to feed contention. The key to understanding policy responses to messy issues, then, lies in discovering how intractable policy controversies come about.

This chapter sifts through recent thinking in the policy sciences to assemble a vocabulary and grammar to talk about intractable policy controversies. This chapter looks for terms to understand where policy conflict about messy issues takes place, what this conflict is about, and who engages in intractable policy controversies. Once the vocabulary is in place, the chapter draws on work inspired by the late Dame Mary Douglas to provide a 'grammar' for explaining persistent conflict about messy challenges (Douglas, 1970; Thompson et al, 1990; Thompson et al, 1999).

The following is not a comprehensive review of the literature in this field.[1] Neither is it an attempt to design a new theory of messy policy problems. Rather, this chapter cobbles together a rough-and-ready framework for analysing intractable policy controversies from a range of existing concepts. There can be no doubt that these ideas and concepts are somewhat incompatible. However, rather than focusing on these incompatibilities, this chapter explores the many synergies between contemporary accounts of the policy process.

Where does conflict about messy problems take place? The differentiated polity and its potential for controversy

Part of the close relationship between messy issues and persistent conflict has to do with the institutional environments in which policy is made today. Over the past three or four decades, the institutional settings for policy-making have changed as rapidly and as profoundly as have our society. In the not too distant past, policy was something produced and owned by 'government'. Working from within recognizable institutions, governments steered societies by making and enforcing rules. Today we use the far more amorphous term 'governance' to describe a confusing myriad of crisscrossing activities, institutions and processes that all seem, in some way or another, to contribute to similarly opaque things called 'policies' (Pierre and Peters, 2000).

Notice the plural! Governance is about making *policies* for a mind-boggling variety of issues and activities. Over the past 30 or so years, states have become involved in an ever-widening scope of activities ranging from economic life to arts or even sports. New demands on policy-making have spawned specialist organizations to deal with them. In the UK, for example, arts policy emerges from a group of organizations including the central Department for Culture, Media and Sport, the Arts Councils for England, Scotland and Wales, local authorities, interest representation (such as British Actors' Equity or the Scottish Artists Union) as well as the myriad of museums, theatres and opera houses around the country.

This expansion has profoundly changed the organizational make-up of the state. Governments used to consist of a few organizations designed for the generic task of governing (rule-making, implementing and sanctioning) in an institutional space clearly labelled 'the state'. Governance systems, in turn, distribute organizational clusters that specialize in a particular issue throughout society. Thus, the state changes from being a monolithic entity – the one we used to call 'government' – to being a collection of institutional networks. Each of these networks is geared towards handling a limited number of policy issues. For example, no one would expect (or even much appreciate, let alone listen to) the Arts Council of England to formulate a policy position on anti-terror measures. By the same token, pundits would probably be surprised to read a White Paper on innovative ways of funding New Music in the UK authored by the Ministry of Defence. In this sense, the term 'governance' describes the institutional fragmentation of the state into specialized areas of competence (Richardson, 1996; Rhodes, 1997). There are as many different terms for these institutional networks as there are researchers involved in studying them (Nullmeier and Rüb, 1993). Common names are 'policy networks' (Rhodes, 1990, 1997), 'policy communities' (Richardson and Jordan, 1987), or 'policy subsystems' (Sabatier and Jenkins-Smith, 1993, 1999).

Governance also implies that the level of engagement with each item on the growing list of challenges becomes more intensive. Now that government has dealt with all the easy problems, Rittel and Webber (1973) note, we are left with the complex and uncertain challenges. Formulating and implementing policy for these types of 'wicked' problems requires specialized knowledge not readily available in-house in the relevant ministries or executive agencies (Atkinson and Coleman, 1992; Adler and Haas, 1992). Rather, specialized technical knowledge can be anywhere within an issue area: professionals, NGOs, lobby groups, firms or citizens may be a source of technical competence for any given problem. For example, the National Autistic Society in the UK not only '... champions the rights and interests of all people with autism ...', it also is a reservoir for specialist knowledge about people with autism spectrum disorders (National Autistic Society, 2008). Similarly, in Germany the *Technischer Überwachungsverein* (TÜV) is an association of technical consultants who '... validate the safety of products of all kinds to protect humans and the environment against hazards' (TÜV, 2008). Policy-makers in Germany have come to rely on the TÜV as a source of knowledge on the risks of new technology. States need to tap this source of specialized technical expertise either by recruiting experts or by inviting interest groups into the policy-making process.

Widening and deepening the remit of policy-making has meant that governance involves more, and a rather different mix of, people than did government. Not only do states need to acquire technical knowledge, they also have to (at least in democratic polities, but also in less democratic ones) take into account the differing interests and political demands that coalesce around a policy issue (Parsons, 1995). Policy-making, then, becomes a process of exchange, transaction and bargaining between different institutions and policy actors within these discrete institutional networks.

Since policy networks are only tenuously connected to each other, the patterns of interaction between actors evolve differently for each policy community. As a result, each network features a specific constellation of organizational and interpersonal relations (Rhodes, 1990; Richardson, 1996). Whereas relations within traditional governments were typically characterized by hierarchical relationships, what political scientist Chris Ansell (2000) calls the 'networked polity' features heterarchical relationships. Rather than mapping many actors to a single central actor – the state – governance features a 'many-to-many' mapping of policy actors. The precise maps of individual and organizational relations differ from network to network and over time.

The debate about 'governance' – what it is, what it does and how it works – is far from settled (Rhodes, 1997; Pierre and Peters, 2000). Commentators will probably continue to argue about the phenomena that constitute governance long after the term itself has become unfashionable. What does seem clear though is that the institutional contexts for policy-making have become more variegated, less coherent and more spe-cialized. R. A. W. Rhodes speaks of the 'differentiated polity' which '… is characterized by functional and institutional specialization and the frag-mentation of policies and politics' (Rhodes, 1997, p7). By widening the scope of policy actors and weakening the hierarchical control of central governments, the differentiated polity creates the potential for intractable policy controversy. On the one hand, governance involves more and more diverse people; on the other hand, since states rely on the expertise of these diverse actors, they can no longer simply overrule or ignore disagreement.

What is policy conflict about?
Making sense of uncertainty

While conflict in a field of actors competing and cooperating with each other is probably inevitable, this does not mean that it need be persistent. What, then, is going on in these policy networks and policy subsystems?

Policy conflicts about messy problems are persistent and intractable because of their inherent uncertainty. We simply do not know for sure how climate change will affect, say, agricultural production, or how demo-graphic ageing will shape demand for health care services, or what effects we can expect transport congestion to have for economic growth in Europe.

This is not to say that we know nothing about these problems. On the contrary, issues such as climate change, transport policy, pension reform or health care provision generate an abundance, some might even argue an overabundance, of research. The availability of data, facts and evidence is not the problem. It is knowing what this data means for policy that proves remarkably difficult and controversial.

And yet, despite considerable uncertainties, policy-makers are called upon to act. Nowadays policy-makers cannot simply act on a whim (at least they cannot be seen to be acting on a whim). Neither can they base their actions on astrology, palm reading or the private revelations of a deity (or at least they cannot be seen to base their decisions on such things). Today, we demand that policy-makers have credible reasons for doing everything. That is why they need to make sense of the stream of facts, data and evidence that tears past them.

The problem is that, contrary to common wisdom, facts never speak for themselves. Most of the time, facts remain stoically silent. If they do speak, they rarely communicate intelligibly. Instead, facts tend to mumble and stutter, rambling on in allegories and metaphors, digressing in counterin-tuitive ways as they point obliquely in this or that direction.

On their own, then, facts are not much use to policy-making. First, the sheer volume of data on any given topic is overwhelming. There are so many facts on almost everything out there that if they were to speak, they would produce a cacophony of mumbling voices. If nothing else, someone or something has to guide policy-makers in selecting which voices to listen to and which voices to ignore. Making sense of these data requires a degree of interpretation and, to prevent being overwhelmed by the mass of data, selection.

Second, policy relevance is not an intrinsic feature of any particular piece of scientific evidence. Science predominantly deals with abstraction, generalizations and universals. Policy is always about a particular problem or condition, itself embedded in specific institutional and situational contexts. Thus, for science to be relevant to policy it needs to engage with particular socio-economic, cultural and political settings. Take climate change for example. In 2004, cosmologists hypothesized that the solar system's gas giant was going through a period of planetary warming (Marcus, 2004). Some commentators started wondering aloud what this meant for climate change policy here on Earth. Yet, this was completely unclear. Did it mean that global warming is unrelated to anthropogenic carbon dioxide emissions? Will carbon dioxide emissions interact with planetary warming or will it have no effect? Does this mean that we scrap the global climate change regime? And, anyway, how should we respond to non-anthropogenic planetary warming? Is it something to worry about at all? These (and many more) questions remain largely unanswered to date. Policy-makers know that they ignore these contextual factors at considerable peril. However, the findings of scientific research often relate to these social contexts opaquely: most of the time, the connection between a scientific and the actual policy is neither obvious nor unambiguous. Rather, policy relevance has to be brought about by skilful interpretation.

Third, the act of applying scientific evidence to real policy problems does not take place in an institutional and political vacuum. On the contrary, the analyst is likely to encounter a variety of individuals and organizations with widely divergent beliefs and interests all negotiating the thicket of institutional structures, rules and practices. What is more, this is the place that analysis comes into contact with wider social and cultural norms and expectations. In other words, outside the safety of the ivory tower, analysts and analysis can expect to get a sceptical reception from a potentially hostile audience no matter how good the science is. Analysts therefore need to convince the audience of the salience and importance of the interpretations and selections that they have made. This requires a fair amount of rhetorical and persuasive skill.

Thus, the policy process needs far more than scientific facts. To be of any use at all, analysis needs to be able to help policy actors answer what Fox and Miller (1995) consider to be the central question in policy-making: What shall we do next? This means situating scientific facts within specific policy contexts. Policy-relevant analysis needs to make the connection between the abstractions of science and the grubby realities of the social world in which these facts are supposed to make a difference. Significantly, they also show policy-makers what these particular facts mean, why policy-makers should pay attention to this rather than another set of data, how these facts are relevant to the particular issues and problems that concern policy-makers and, most importantly, what should be done about the problem in question (Stone, 1988; Majone, 1989; Fischer and Forester, 1993). In short, good policy analysis relies on selection, interpretation and persuasion to transform facts into compelling arguments.

Fashioning arguments requires judgement. What, then, guides judgement? What criteria do policy actors use to select and interpret data?

Policy scientists studying these processes are convinced that ideas, beliefs and world-views inform judgement about messy challenges (Thompson and Schwarz, 1990; Rayner, 1991; Haas, 1992a). By selectively highlighting some aspects while backgrounding others, world-views help policy actors to identify salience and relevance of data for policy-making. Crucially, the criteria for fore- and back-grounding are 'trans-scientific' (Weinberg, 1972): they are beyond validation by rational, objective or scientific means. For this reason, interpretation of messy policy problems relies on values that make up different world-views (Thompson and Tayler, 1985; Rayner, 1991). Judgement and, ultimately, analysis guided by different world-views invariably generates divergent interpretations on the same policy issue. Thus, incommensurable values of different world-views give rise to conflict in policy debates.

A convenient shorthand for depicting world-views and the way they shape policy processes is the idea of a 'frame'. Rein and Schön (1994) argue that applying frames to social reality 'is a way of selecting, organizing, interpreting, and making sense of a complex reality to provide guideposts for knowing, analysing, persuading, and acting. A frame is a perspective from which an amorphous, ill-defined, problematic situation can be made sense of and acted on.' (p146)

Frames give policy actors the cognitive and normative tools to interpret events, facts and scientific evidence. Frames allow policy actors to ascribe meaning and significance to political life. These frames cannot, argue Rein and Schön, be reduced to interests or preferences because '... it is the frames held by the actors that determine what they see as being in their interests and, therefore, what interests they perceive as conflicting' (Rein and Schön, 1994, p29, original emphasis). In this sense, frames cannot be falsified: Rein and Schön observe that '... if objective means frame-neutral, there are no objective observers' (Rein and Schön, 1994, p30). Policy arguments, then, are based in judgement guided by one or another frame. Box 2.1 provides examples of different types of frames.

Box 2.1 *Dryzek's five frames*

John Dryzek identifies five social science frames relevant for policy analysis: welfare economics, public choice theory, information processing, social structure and political philosophy. The existence of multiple frames, he argues, undermines the feasibility of objective and uncontested scientific truths. In real policy processes, he contends,

> ...[n]umerous social science frames of reference can be applied to the analysis of policy. It is not just that these frames give different answers to policy questions. Rather, each frame treats some topics as more salient than others, defines social problems in a unique fashion, commits itself to particular value judgements, and generally interprets the world in its own particular and partial way. (Dryzek, 1993, p222)

Here, frames are similar to 'a language or even a culture shared by a tribe of experts'. As with frames in Rein and Schön's approach, Dryzek argues that policy actors use frames to construct multiple and conflicting interpretations of social reality, not least the reality of complex and uncertain policy issues. The five frames, he points out, '... show that multiple theories can be brought to bear in any given situation. And each frame comes complete with both a lens for interpreting the world and procedures for testing its own hypothesis – but not necessarily for testing those generated in other frames.' (Dryzek, 1993, p223)

This, then, is what causes conflict about messy policy problems to be intractable. Contending frames generate incompatible arguments about 'what we should do next'. Any particular framing of a messy policy issue is likely to be contested by people who use another frame. Thus, debate about messy issues is likely to develop into a controversy played out in terms of these policy arguments. In such an argumentative policy process, controversies ...

> ... *cannot be understood in terms of the familiar separation of questions of value from questions of fact, for the participants construct the*

problems of their problematic policy situations through frames in which facts, values, theories and interests are integrated. Given the multiple social realities created by conflicting frames, the participants disagree both with one another and also about the nature of their disagreements. (Rein and Schön, 1993, p145)

We can now see why this conflict is so stubbornly persistent. Conflict in an argumentative policy process is expressed using the language of science, fact and evidence. The origin of this conflict, however, lies in the incompatible values that guide the judgement required in making any sense of messy policy issues. And yet, firm in the belief that facts can resolve the debate one way or another, policy actors scream for more evidence. The new facts remain as taciturn as the old facts. Policy actors need to use their frame-based judgement to make sense of these facts and fashion them into policy arguments. In an attempt to strengthen the legitimacy of their policy position, parties to a controversy will pit these policy arguments against one another in the policy subsystem. This, leads to a renewed round of controversy. So, paradoxically, far from resolving policy conflict, objective evidence actually fans the flame of policy contention.

Who takes part in conflict about messy issues? Policy actors, coalitions and their stories

Of course, arguments and accounts of social reality, no matter how conflicting they may be, do not really vie for legitimacy in the partial public sphere of policy subsystems. In fact, arguments do not do much of anything. Arguments are just ideas expressed in words. Just like the guns that are not purported to kill people, arguments are, in themselves, harmless ... unless, that is, someone picks them up, carries them into the public sphere and fires them at someone else.

It is people that, sadly much like the killing, do all of the picking up, carrying and hurling of arguments. But, of course, given the size and complexity of contemporary policy processes, not all people are involved in the business of picking up, carrying and hurling policy arguments. Indeed, the differentiated polity would prevent them even if they wanted to. The sheer number of policy networks, most of which are highly specialized, is likely to overwhelm the most educated of citizens.

So who does all of this picking up, carrying and hurling? Put a little more technically, how can we think of the conflicting policy actors in argumentative processes that take place in policy subsystems?

For a long time, the social sciences worked with a rather reduced model of a policy actor. According to this model, the best way to understand policy-making is to assume that policy actors are essentially rational. They know what they want and can rank these wants. They use these wants – the technical term is preferences – to guide their action and interaction with other, equally rational, people in the policy process. On this view, persons involved in policy-making try to get most of what they want out of the policy process. In order to do so, they will strategically use any resource at their disposal. This includes things such as money, influence or skill. It also includes knowledge and experience. For example, knowing the way political and legal institutions work is an invaluable asset for furthering one's own agenda or obstructing any other people's agendas perceived to be a threat.

The rational actors' model also assumes that there is no limit to the quality and quantity of wants. Because people are essentially individuals, they want all sorts of different things and they want to get as much as they can of them. At the same time, so the argument goes, there are not enough resources for everyone to satisfy their wants. Hence, different rational individuals need to compete for scarce resources. For policy-making, this means that rational individuals compete for access to political power to realize their objectives. According to this model, then, policy emerges from the competition between rational individuals.

In mass societies with a wide range of organizations, political scientists merely expanded the idea of a rational decision-maker to organizations. Each organization, it was argued, is just like a rational individual. Unitary organizations – as they were called – have policy objectives (wants) they pursue by rationally deploying the resources available to them. In policy processes with many rational actors, the model predicted that policy-making is an incremental process (Lindblom, 1958). When contending policy actors clash and one cannot easily dominate the other, they look for policy solutions that help to get both of them a little of what they wanted.[2] Political scientists assumed that this process of compromise and adaptation would resolve the conflict about who gets what. Let's design a policy on X, the hypothetical negotiation would run, that helps both of us as much as possible without either of us having to unduly make sacrifices. Or, alternatively, you get more this time around but next time it will be my turn to get more.

In general, so political scientists of the time argued, incremental politics militate against large-scale changes. And a good thing too, they added. In political systems that allowed many rational actors to have a real voice in policy-making, competition would ensure that all organizations got some of what they wanted some of the time but that no organization could get all it wanted all of the time. Policy outcomes were fairly balanced – meaning good some of the time and not so good the rest of the time – for all participants. This, essentially, is how the political scientist Robert Dahl (1962, 1971) defined political pluralism.

Rivers of ink have been spilled about the shortcomings of the rational actor model. Sociologists from Durkheim onwards, social anthropologists such as Mary Douglas, philosophers such as Jürgen Habermas, psychologists such as Daniel Kahneman, and even some economists (Josef Schumpeter comes to mind) have found many reasons to be sceptical of the rational actor model. In the policy sciences, people such as Deborah Stone (1988), Gianfranco Majone (1989), Frank Fischer (1993, 2003) and John Dryzek (1993) have systematically taken apart the notion of rational policy actors by looking closely at the role of analysis and science in the policy process.

For our purposes, the rational actor model is simply impractical. It tells us little about the way policy actors deal with messy policy issues. In order to get what you want, it helps to know what you want. The rational actor paradigm needs to assume that actors have well-formed preferences. But conflict about messy problems is precisely about figuring out what we want from the climate change regime, pension systems or health policy. Debate in an argumentative policy process is as much, if not more, about negotiating meaning as it is about determining who gets what.

While the incrementalism assumes that actors agree on the stakes, messy problems expand the conflict to the very base of policy-making (Wildavsky, 1987; Thompson and Ney, 1999). What kind of policy issue is climate change? Is it an issue of poverty and inequality, as many of the developing countries claim? Is it an issue of technology? Is it an issue of control and regulation? Is it an issue of profligate consumption? Or, is it all of these things? If so, how do they interact? This is why conflict is intractable: policy actors cannot actually agree what is at issue, let alone negotiate a compromise between different preferences. In an argumentative policy process, policy actors need to learn as they try to make sense of messy policy issues. So what we want is likely to change as we learn. Yet, what we learn, in turn, will depend on what we want.

For these reasons, understanding conflict about messy issues requires a different model of the policy actor. This model must come to grips with the argumentative policy process. It needs to help us understand how policy actors use knowledge and ideas – frames – to make sense of messy policy problems in the differentiated polity.

Luckily, policy scientists such as Peter Haas (1992a), Paul Sabatier and Hank Jenkins-Smith (1993, 1999) and Maarten Hajer (1993) have developed a range of similar models that fit the bill nicely. In an argumentative policy process, these theorists suggest, individuals coalesce into groups around a particular frame. On this view, unlike the rational actor model, the motivation for policy action is no longer internal to the individual person or organization. Rather, individuals draw their motivation for policy action from shared norms and practices. In an argumentative policy process, then, the focus of our analytical attention is a group of individuals that makes sense of and acts on messy policy issues in terms of a shared frame.

Depending on the particular model, the composition of the group can differ. All models, however, imply that, since organizations have become large and complex social spaces, commitment to a particular frame is a stronger motivation than loyalty to a particular institution (Sabatier and Jenkins-Smith, 1993). Or rather, these models assume that policy actors express their loyalty to particular institutions by operating in coalitions and communities held together by a frame. On this view, unitary organizations are the exception rather than the rule.

Some groups may consist of people who share a profession, such as scientists or doctors. Peter Haas (1992a) calls these groups 'epistemic communities'. They consist of scientists or other professionals who think about and exercise their profession in ways that sets them off from their colleagues. These shared norms and practices also tie the members of an epistemic community into a common policy enterprise. By way of example, Peter Haas points to the group of scientists who believed that the data on stratospheric ozone showed that CFCs were responsible for the destruction of the ozone layer (Haas, 1992b). It is these epistemic communities, Haas argues, that help policy-makers make sense of technically complex policy issues such as stratospheric ozone depletion.

Other thinkers see policy communities emerging across both organizational and professional boundaries. Maarten Hajer (1993) suggests that policy actors form what he likes to call 'discourse coalitions'. A coalition, he continues, consists of '... a group of actors who share a social construct' (Hajer, 1993, p43). Similarly, Paul Sabatier and Hank Jenkins-Smith (1993, 1999) argue that policy results from the argumentative struggle between what they call 'advocacy coalitions' in policy subsystems. An advocacy coalition consists of ...

> *... people from a variety of positions (elected and agency officials, interest group leaders, researchers, etc.) who share a particular belief system – that is, a set of basic values, causal assumptions, and problem perceptions – and who show a non-trivial degree of co-ordinated activity over time. (Sabatier and Jenkins-Smith, 1993, p25)*

For our purposes, both Hajer's 'discourse coalitions' and Sabatier and Jenkins-Smith's 'advocacy coalitions' describe very similar phenomena.[3] In both approaches, individuals form groups around what are, to all intents and purposes, frames. Sabatier and Jenkins-Smith call these frames 'policy belief-systems'. Hajer calls these frames 'discourses'. In either approach, individuals use shared ideas, norms and practices to make sense of messy challenges and thereby influence policy.

On this view, it is advocacy coalitions[4] that do all the assembling, lugging and hurling of policy arguments. In the differentiated polity, these communities of like-minded people use common frames to interpret messy problems. The frames help them to impose some sort of cognitive order on the complexity and uncertainty of issues such as climate change or the global health crisis. They help them recognize and formulate their preferences. Most importantly, they help them to formulate persuasive and coherent arguments for or against a particular course of action.

How, then, can we get a handle on the assembling, lugging and hurling of policy arguments?

One way of thinking about this process of public cogitation is in terms of policy stories or policy narratives. Policy stories really are nothing but handy metaphors for describing the things that go on within and between advocacy coalitions. On this approach, advocacy coalitions cast their coherent and persuasive policy arguments in the form of narrative accounts of social problems. Rein and Schön argue that '... these problem-setting stories, frequently based on generative metaphors, link causal accounts of policy problems to particular proposals for action and facilitate the normative leap from 'is' to 'ought' (Rein and Schön, 1993, p148). Policy arguments, then, tell a story: they outline a setting, provide heroes and villains, suggest a solution and, most importantly, are guided by a moral (Stone, 1997).

Advocacy coalitions use stories to 'deliver' policy arguments to a critical, possibly hostile but definitely sceptical audience. By looking at policy issues through the 'perceptual lenses' (Allison, 1971) of a particular frame and by packaging these findings in a narrative format, policy actors try to win over others to their view of what is going on. Stories enable advocacy coalitions to communicate a particular argument by telling others what the problem is, who or what is responsible for causing it, and how to go about solving the problem. Policy stories, Stone concludes,

> *... are tools of strategy. Policy makers as well as interest groups often create problems (in the artistic sense) as a context for the actions they want to take. This is not to say that they actually cause harm and*

destruction so they will have something to do, but that they represent the world in such a way as to make themselves, their skill, and their favourite course of action necessary. (Stone, 1997, p162)

So, what do we know about the way policy actors deal with complex, uncertain and transversal policy problems? First, we know that, in the differentiated polity, policy-making takes place in discrete institutional networks focused on a particular policy problem. Second, we have also seen that conflict in these discrete institutional networks – interchangeably called policy networks, policy communities or policy subsystems – is what Rein and Schön (1993, 1994) have called an 'intractable policy controversy'. Conflict is intractable because policy problems are uncertain and complex. Making sense of these problems requires skilful use of selection, interpretation and rhetoric. Choosing carefully from the available wealth of data, policy actors will fashion facts into coherent and persuasive arguments that say something meaningful about the policy context at hand. This process involves judgement which itself is guided by the ideas that make up frames, since frames encode values, and competing frames generate a wide range of meaningful yet conflicting things to say about a policy problem. Scientific evidence, far from resolving conflict, fans the flames of contention because it is refracted through the 'perceptual lenses' of contending frames.

Third, we also know that much of this interpretation and argumentation is done by individuals who share a frame and who coordinate their action. Policy scientists call these groups discourse coalitions or advocacy coalitions. In debates about messy challenges, advocacy coalitions apply their respective frames to develop persuasive arguments. They communicate these arguments in the form of narratives or stories to mobilize support for a particular course of action. But because frames are based on fundamentally conflicting values, the narratives invariably tell conflicting stories about messy issues. In short, conflict about messy issues is persistent and intractable because it is based on contending frames.

So far so good. But where do frames come from and what makes them differ? Is there a way of comparing different frames and the policy stories they generate? How do different coalitions interact in policy subsystems? Is there a way of identifying the patterns of agreement and disagreement between policy stories? And, most importantly, can we predict the impact of frames and stories on policy?

Luckily, there is and we can.

A typology of coalitions, frames and stories

In order to analyse frame-based policy conflict about messy issues, we need to find a way to relate advocacy coalitions to frames, and frames to policy stories. The approach inspired by the late social anthropologist Dame Mary Douglas – sometimes referred to as 'cultural theory' – provides us with just the grammar to lay out our vocabulary into a useful explanation of how policy-makers deal with messy issues.

Cultural theory is based on the simple but powerful insight that the way we organize our social relations shapes the way we perceive the world and, consequently, the way we behave. 'Cultural theorists' look to '... the various ways in which we bind ourselves to one another' or 'social solidarities' to explain individual and collective behaviour (Thompson et al, 1999). Social solidarities, Thompson, Grenstad and Selle argue, consist of distinctive but coherent patterns of social relations, accompanying frames (which cultural theorists call 'cultural biases'), and characteristic patterns of behaviour (Thompson et al, 1999). These three elements reproduce and support each other in complex and non-linear ways. Social structures shape the cultural biases or frames that individuals use to socially construct the world. These legitimate specific patterns of behaviour which, in turn, reinforce and reproduce social structures and world-views.

Cultural theory, enables analysts to map social solidarities and analyse the complex social dynamics that result from their interaction. However, cultural theory is not about measuring or ranking social solidarities against an external standard of objectivity or rationality. In a world of plural social solidarities, cultural theorists know that knowledge and truth are necessarily relative and socially constructed. Yet, unlike with much of post-modern social and political thought, cultural theorists see no reason why there should be an infinite number of ways in which policy actors can socially construct the world. Rather, socio-institutional relativism is inherently constrained: at any one time, cultural theorists argue, social systems consist of no more than five competing social solidarities.[5]

Box 2.2 *Grid and group*

Although the 'impossibility theorem' (Schmutzer and Bandler, 1980; Thompson et al, 1990) provides the rigorous and formal foundation for cultural theory today, the basic idea of contesting plural social solidarities emerged from a more pragmatic context (Thompson et al, 1999). Suspecting there to be a 'concordance between symbolic and social experience', the social anthropologist Mary Douglas needed a scheme to identify and classify social contexts (Douglas, 1970; Douglas, 1982). She suggested mapping and comparing the '... interaction of individuals within two social dimensions. One is order, classification, the symbolic systems. The other is pressure, the experience of having no option but to consent to the overwhelming demands of other people.' (Douglas, 1970, p81)

These two dimensions – 'grid' and 'group' respectively – make up the horizontal and vertical axes of a system of coordinates. The vertical grid axis depicts '... the complementary bundle of constraints on social interaction, a composite index of the extent to which people's behaviour is constrained by role differentiation, whether within or without membership of a group.' (Rayner and Gross, 1985, p6) The horizontal group axis, in turn, '... represents the extent to which people are restricted in thought and action by their commitment to a social unit larger than the individual.' (Rayner and Gross, 1985, p5)

As we move along the group dimension, '... the individual is more and more deeply committed to a group, so choices are more standardized as we move from the left across the diagram.' (Douglas, 1996, p68) Moving up the grid dimension depicts social contexts in which discriminating systems of rules become increasingly dense and complex. Thus we should expect a high grid score 'whenever roles are distributed on the basis of explicit public social classifications, such as sex, colour, position in a hierarchy, holding a bureaucratic office, descent in a senior clan or lineage, or point of progression through an age-grade system.' (Rayner and Gross, 1985, p6)

Taken together, grid and group generate a 2×2 matrix or map of social solidarities (see Figure 2.1).

The scope of policy conflict: five social solidarities

Two of the five social solidarities on the cultural map (see Figure 2.1), individualism and hierarchy, are well known to the social sciences. Though equally important for the viability of social systems, the other three cultural archetypes – egalitarianism, fatalism and autonomy – are relatively unknown to social research and have not been systematized with respect to one another as, for instance, is the case with individualism and hierarchy (cf. Williams, 1975).

Members of the individualist social solidarity prefer ego-centred networks that allow for maximum individual spatial and social mobility. Since these networks feature weak social controls in the form of rules, tradition or customs, individuals regulate their relations with other free agents through negotiation and contract. For this reason, members of the individualist social solidarity champion individual rights, liberties and responsibilities. The world, they argue, is there for the taking; what may

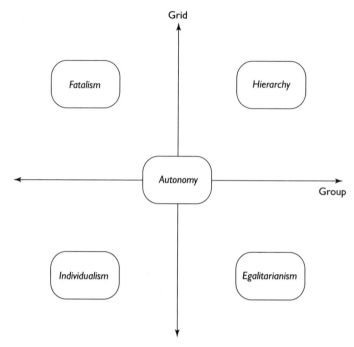

Figure 2.1 The cultural map

seem to others to be problems or barriers are simply opportunities for innovative and hard-working entrepreneurs. For individualists, human nature, characterized by a healthy dose of self-interest, is robust and innovative, and positively thrives on challenges. Here, competition with others for scarce resources is the strategy of choice, and what counts is the bottom-line. The best environment for human development, so the individualist argument goes, is the rough and rugged world of the free market. Since the market rewards intelligence, acumen and hard work, while punishing free-riding, individuals either innovate or suffer the consequences of failure (Douglas and Ney, 1998).

The hierarchical social solidarity, by contrast, is characteristic of '... tradition-bound institutions in which everyone knows his place, but in which that place might vary with time' (Rayner and Gross, 1985). Rather than individual rights and liberties, members of 'nested bounded groups' (Thompson and Schwarz, 1990) emphasize duty, obligation and loyalty toward the institution, be it a tribe, a ministry or a regiment. Hierarchies come equipped with a wide range of explicit and implicit social control mechanisms such as rules and regulations, traditions and customs, as well as ancestry and lineage. At each institutional level, a battery of finely tuned and appropriately tailored norms, incentives and sanctions guide individual behaviour, rewarding loyalty and punishing transgression. For hierarchical actors, managing and regulating social relations (or, indeed, anything) is very much a hands-on task and is best organized from the top down. Left to its own devices, so the argument goes, the world would inevitably descend into disorder and chaos. Human nature, weak and fallible as it is, yearns for order and stability. For this reason, forsaking social mobility and competition in return for the security and order of a strong institution is a small price to pay for members of the hierarchical social solidarity.

Egalitarian actors live in networks that do not differentiate between members while clearly distinguishing the group itself from the outside world. Egalitarian social relations feature few, if any, formal rules regulating conduct (unlike hierarchies) and entail the rejection of measures of social distance such as wealth and physical prowess (unlike individualists). Owing to the absence of clear rules and regulations for succession, leadership tends to be charismatic. As a result, egalitarian enclaves are often in a state of mobilization: suspicion of contamination from the outside and treason from within make this form of organization the most unstable and fluid. Yet this endemic fission is not seen as a problem by those committed to the principle of 'small is beautiful'. Applying a critical rationality to the world, egalitarians typically rail '... against formality, pomp and artifice, rejecting authoritarian institutions, preferring simplicity, frankness, intimate friendship and spiritual values' (Thompson, 1996). Humans, egalitarians argue, are fragile and under duress: coercive institutions with their inhumane hierarchies as well as ruthless markets that value everything in terms of money stunt individual development and smother the delicate human personality. Small and intimate societies based on absolute equality and intersubjective solidarity, in turn provide a suitable environment for individual well-being and the satisfaction of real human needs (Douglas and Ney, 1998).

Fatalism describes marginal and precarious social environments. Individualized by social isolation yet still subject to stringent regulations, fatalists lack both the freedom of choice afforded by open social networks and the security of cohesive groups. With little control over their lives and no means of influencing their social or natural environments, fatalists simply concentrate on coping (Thompson and Schwarz, 1990). For fatalists, the world and society are inherently inscrutable and capricious. Other people, even those who share their social isolation, remain 'mysteriously unpredictable' and untrustworthy (Douglas and Ney, 1998). In fact, fatalists trust in nothing at all except the certain knowledge that there is no rhyme or reason, no telos and certainly no justice in life. Bobbing along in the cold ocean of life, being dragged this way and that by strong but unpredictable currents, it is all fatalists can do to keep their chins above the water.

Autonomy is the preferred form of social organization of the hermit. While the other four forms are based on some form of coercion, the realm of autonomy is a space free of all social pressures. Thompson argues that autonomy is '... a curious sort of solidarity ... because it stabilizes itself by the deliberate avoidance of all coercive involvements' (Thompson et al, 1999, p11). Unwilling or unable to compromise autonomy and self-determination by interacting with others, the hermit withdraws completely from society.

Each of the social solidarities comes equipped with its own frame or cultural bias. Each frame justifies and legitimates a fundamentally incompatible way of social organization.

Dynamics of policy conflict: Self-organizing disequilibrium systems

Social solidarities, cultural theorists argue, are in constant rhetorical and symbolic conflict about how to best structure social relations. This conflict erupts and re-erupts because each social solidarity defines itself in contradistinction to the others. Choosing one social solidarity implies rejecting the others. Mary Douglas maintains that any ...

> ... *act of choice is also active in their [the individuals'] constitution-making interests. A choice is an act of allegiance and a protest against the undesired model of society. On this theory each type of culture is by its nature hostile to the other three cultures. Each has its strengths, and in certain circumstances each culture has advantages over the others. And each has its weaknesses. But all four coexist in a state of mutual antagonism in any society at all times. (Douglas, 1996)*

Somewhat against intuition, cultural theorists maintain that the symbolic and rhetorical conflict among social solidarities is a source of social cohesion. As Douglas points out, each cultural bias socially constructs the world by semiotic selection and negation: each cultural view contains blind-spots and weaknesses. Each socio-institutional form, whether it be the individualism, hierarchy, egalitarianism or fatalism, is suited to solving different types of social problems (Thompson et al 1990; Schmutzer, 1994).[6] The weaknesses of one particular social solidarity creates problems that another solidarity is able to solve. Thus, both cognitively and institutionally, the different cultures depend on one another to define and reproduce themselves.

This insight has been developed into the so-called 'requisite variety condition' (Thompson et al, 1990; Thompson, 1997; Thompson et al, 1999).

In order to function at all, the argument goes, any social entity, whether it is a nomadic tribe, a political party, a multinational corporation, or a nation-state, must have all four cultural types engaged in rhetorical and symbolic conflict. Any social system lacking one or more social solidarities will, ultimately, run into serious problems because it cannot rely on the competitive intervention of other cultures to counter-balance the disadvantages of individual cultural archetypes. Although each social solidarity has its specific weaknesses, each is also expert in pointing to the weaknesses of rival social solidarities. For this reason, Thompson argues that social interaction between the different cultures is necessarily a complex, dynamic and chaotic conflict. Social systems containing their full complement of social solidarities are what Thompson calls 'self-organizing disequilibrium systems'. Here, each social solidarity aims to achieve viability in a competitive environment. In doing so, each culture disorganizes other solidarities: social interaction is always conflictual and always detrimental for one or more of the competing cultures. Yet, at the same time, each culture needs the others to be able to define itself in contradistinction to them. Social change in these types of system is ubiquitous but complex and unpredictable. Therefore, cultural theorists argue that social systems are indeterministic, non-linear, far from equilibrium and highly sensitive to initial conditions (Thompson et al, 1999, p10).

In a very real sense, the concept of socio-cultural viability is about understanding the different ways in which social structures relate to specific frames. The typology of social solidarities provides a means of mapping different institutional contexts to different types of cosmologies. What is more, by grounding conflicting frames in incompatible forms of social organization, cultural theory provides a way of dissecting policy conflict about messy issues.

The scope, structure and impact of conflict about messy policy challenges

The grammar of cultural theory allows us to use the new vocabulary to analyse frame-based policy conflict. Although Sabatier and Jenkins-Smith (1993, 1999) say little about the differences between advocacy coalitions, we can imagine that advocacy coalitions evolve into all sorts of organizational forms. Some may be very informal but have a very clear idea of who they are and what they stand for. For example, the anti-globalization movement consists of a very wide range of individuals and organizations, such as *Association pour la Taxation des Transactions pour l'Aide aux Citoyens* (ATTAC) or the People's Health Movement, that have a very clear message but no mechanisms of internal organization. Other advocacy coalitions may be more defined and regimented; these coalitions may even have a name or some formal organization. For example, the Physicians for a National Health Program (PNHP), a group of North American physicians campaigning for a universal health systems in the USA, represents a coalition of professionals with a clear mission, internal structure and boundary.[7] Others still may lack any formal mechanisms of coordination and may consist of ideologically heterogeneous members. These coalitions may consist of short-term alliances focused on a specific issue or task, say preventing the passing of a bill or the siting of a facility. After the issue goes off the boil or the task is completed, the coalition disbands. For example, the initiative 'Manchester Against Road Tolls' (MART) aims to influence the referendum about introducing a road pricing system in Manchester (MART, 2008). This coalition consists of motorists' associations (e.g. RiderConnect or the Association of British Drivers), civil society groups (e.g. the Manchester Civic Society), local firms (e.g. Prestwich Antiques) and political parties (e.g. Tameside Conservatives). The members have little in common except their disdain for congestion charging. Over and above a website, the coalition uses few formal organizational mechanisms to coordinate their activity (MART, 2008).

Cultural theory's typology of social solidarities allows us to model advocacy coalitions and their frames in any particular policy subsystem. On this view, advocacy coalitions organize along the lines suggested by the typology of social solidarities. Policy subsystems feature at least three types of active advocacy coalitions: one organized along hierarchical lines, another featuring looser individualist organization and a third resembling a more egalitarian setup.[8] Each type of advocacy coalition comes with its own basic frame for making sense of messy policy challenges. Members of different types of advocacy coalitions qua social solidarities use these frames to construct narratives about what they think is going on and what they think ought to be going on.

Using policy stories as a means of articulating and comparing narratives, we can dissect and compare intractable conflict about messy issues in three ways:

First, we will apply our new language to understand the *scope* of policy conflict about messy policy challenges. Since advocacy coalitions use frames grounded in fundamentally divergent forms of social organization, we should expect conflict between contending advocacy coalitions. Since we are more interested in comparing the content of policy stories rather than the form, the studies compare policy stories in terms of their basic assumptions (the setting), the way they define the problem (the villain) and the preferred solution to the problem (the heroes). The aim is to show how incompatible frames, based in a particular mode of social organization, give rise to contradictory accounts of the same messy policy challenge. In this way, the analysis can gauge the distance between different advocacy coalitions and the policy stories they tell.

Second, the language of policy conflict also allows us to compare the *structure* of conflict about messy policy challenges. As we have seen, conflict between contending advocacy coalitions is the norm in argumentative policy processes. Each advocacy coalition will define itself and its policies in contradistinction to the other types of advocacy coalitions in the policy subsystem. However, this is not to say that there is no agreement at all. Cultural theory suggests that pairwise alliances between two advocacy coalitions emerge along the grid and group dimensions. The typology of frames helps unearth the areas of agreement and disagreement between contending advocacy coalitions. Advocacy coalitions will agree on specific interpretations and suggested policy solutions. More importantly, advocacy coalitions will also agree on the types of arguments they reject. In this way, the analysis can explore potential avenues of cooperation between advocacy coalitions.

Last, the analysis also allows us to project the potential *impacts* of advocacy coalitions' policy arguments. Since each frame highlights certain issues and backgrounds others, contending advocacy coalitions are good at spotting and rectifying some aspects of messy policy issues. However, given the selective nature of frame-based judgement, advocacy coalitions are less perceptive to other, equally important, dimensions of messy problems. By contrasting respectively the strengths and weaknesses encoded in each policy story, the analysis can predict some of the 'unexpected'[9] consequences of proposed solutions.

This three-pronged analysis, then, enables us to map conflict about messy policy problems. The scope of policy conflict marks the outer boundaries of an argumentative space – the contested terrain of policy-making. The three policy stories delimit the set of ideas, facts and methods of the policy debate. The structure of policy conflict represents the topography of conflict about messy issues within the argumentative space. The areas of agreement and disagreement between contending advocacy coalitions define the shape of the landscape in which deliberation about a particular messy policy challenge takes place. The potential impact of policy conflict points towards movement on the landscape as well as the evolution of the policy space itself.

Notes

1 Luckily, Frank Fischer (2003) has already written that particular book.
2 The approach assumed that rational actors would prefer a little of what they wanted over nothing at all.
3 In his comprehensive review of approaches to policy-making that use discourses and ideas, Frank Fischer (2003) devotes an entire chapter to comparing discourse coalitions with the Advocacy Coalition Framework. He comes down heavily on the side of Maarten Hajer's concepts because, so the argument goes, Sabatier and Jenkins-Smith have overly simplified the role of discourse. In their attempt to create an empirically grounded set of concepts, Fischer argues, the two policy scientists have lost the ability to appreciate the 'transformative' aspects of discourse.
4 Despite its weaknesses, I prefer Sabatier and Jenkins-Smith's term; it seems to describe far better what policy actors do.
5 Technically, this is what Michael Thompson calls the 'impossibility theorem'. For a mathematical treatment of the impossibility theorem, see Schmutzer and Bandler (1980) and Schmutzer (1994).
6 The autonomy of the hermit by definition solves no problems of other social solidarities.
7 In many respects, the PNHP is closer to an epistemic community than an advocacy coalition.
8 Why only three of the five social solidarities? Cultural theorists argue that only the hierarchical, individualist and egalitarian solidarities participate actively in policy processes. Fatalists, owing to their inherent social isolation, neither have the means nor see any sense in getting involved in politics. The hermit, by definition, avoids social commitments of any sort.
9 Unexpected, at least, for proponents of the particular policy story.

References

Adams, J. (1995) *Risk: the Policy Implications of Risk Compensation and Plural Rationalities*, UCL Press, London
Adler, E. and Haas, P. (1992) 'Conclusion: Epistemic Communities, World Order, and the Creation of a Reflective Research Program', *International organization*, vol 46, no 1, pp367–389
Allison, G. (1971) *The Essence of Decision: Explaining the Cuban Missile Crisis*, Little Brown, Cambridge, MA
Ansell, C. (2000) 'The Networked Polity: Regional Development in Western Europe', *Governance: An International Journal of Policy and Administration*, vol 13, no 3, pp303–333
Atkinson, M. and Coleman, W.D. (1992) 'Policy Networks, Policy Communities and the Problem of Governance', *Governance: An International Journal of Policy and Administration*, vol 5, no 2, pp154–180
Dahl, R. (1962) *Who Governs? Democracy and Power in an American City*, Yale University Press, New Haven, CT
Dahl, R. (1971) *Polyarchy: Participation and Opposition*, Yale University Press, New Haven, CT

Douglas, M. (1970) *Natural Symbols: Explorations in Cosmologies*, The Cresset Press, London

Douglas, M. (1996) *Thought Styles: Critical Essays on Good Taste*, Sage Publications, London

Douglas, M. (ed) (1982) *Essays in the Sociology of Perception*, Routledge and Kegan Paul, London

Douglas, M. and Ney, S. (1998) *Missing Persons: A Critique of Personhood in the Social Sciences*, University of California Press, Berkeley, CA

Dryzek, J. (1993) 'Policy Analysis and Planning: from Science to Arguments', in F. Fischer and J. Forester (eds), *The Argumentative Turn in Policy Analysis and Planning*, Duke University Press, Durham, NC

Fischer, F. (2003) *Reframing Public Policy: Discursive Politics and Deliberative Practices*, Oxford University Press, Oxford

Fischer, F. and Forester, J. (eds) (1993) *The Argumentative Turn in Policy Analysis and Planning*, Duke University Press, Durham, NC

Fox, C. J. and Miller, H. T. (1995), *Postmodern Public Administration: Towards Discourse*, Sage Publications, Thousand Oaks, CA

Haas, P. (1992a) 'Introduction', *International organization*, vol 46, no 1, pp1–35

Haas, P. (1992b) 'Banning Chlorofluorocarbon: Epistemic Community Efforts to Protect Stratospheric Ozone, *International organization*, vol 46, no 1, pp186–224

Hajer, M (1993) 'Discourse Coalitions and the Institutionalization of Practice', in F. Fischer and J. Forester (eds) *The Argumentative Turn in Policy Analysis and Planning*, Duke University Press, Durham, NC

Lindblom, C. (1958) 'Policy Analysis', *American Economic Review*, vol. 19, pp78–88

Majone, G. (1989) *Evidence, Argument and Persuasion in the Policy Process*, Yale University Press, New Haven, CT

Marcus, P. (2004) 'Prediction of a Global Climate Change on Jupiter', *Nature*, vol 428, no 6985, pp828–831

MART (2008) 'About MART', www.manchestertolltax.com/index.php/mart-aims, accessed 23 October 2008

National Autistic Society (2008) 'Welcome', www.nas.org.uk, accessed 16 October 2008

Nullmeier, F. and Rüb, F. (1993) *Die Transformation der Sozialpolitik: Vom Sozialstaat zum Sicherungsstaat*, Campus Verlag, Frankfurt, a.M.

Parsons, D. W. (1995) *Public Policy: an Introduction to the Theory and Practice of Policy Analysis*, Edward Elgar, Aldershot

Pierre, J and Peters, B. G. (2000) *Governance, Politics, and the State*, St Martin's Press, New York, NY

Rayner, S (1991) 'A Cultural Perspective on the Structure and Implementation of Global Environmental Agreements', *Evaluation Review*, vol 15, no 1, pp75–102

Rayner, S. and Gross, J. (1985) *Measuring Culture: a Paradigm for the Analysis of Social organization*, Columbia University Press, New York, NY

Rein, M. and Schön, D. (1993) 'Reframing Policy Discourse', in F. Fischer and J. Forester (eds) *The Argumentative Turn in Policy Analysis and Planning*, Duke University Press, Durham, NC

Rein, M and Schön, D. (1994) *Frame Reflection: Towards the Resolution of Intractable Policy Controversies*, Basic Books, New York, NY

Rhodes, R. A. W. (1990) 'Policy Networks: A British Perspective', *Journal of Theoretical Politics*, vol 2, no 3, pp293–317

Rhodes, R. A. W. (1997) *Understanding Governance: Policy Networks, Governance, Reflexivity and Accountability*, Open University Press, Buckingham

Richardson, J. J. (ed.) (1996) *European Union: Power and Policy-Making*, Routledge, London

Richardson, J. J. and Jordan, A. G. (1987) *British Politics and the Policy Process: An Arena Approach*, Allen & Unwin, London

Rittel, H. and Webber, M. (1973) 'Dilemmas in a General Theory of Planning', *Policy Sciences*, vol 4, pp155–169

Sabatier, P. and Jenkins-Smith, H. (1993) *Policy Change and Learning: An Advocacy Coalition Approach*, Westview Press, Boulder, CO

Sabatier, P. and Jenkins-Smith, H. (eds) (1999) *Theories of the Policy Process*, Westview Press, Boulder, CO

Schmutzer, M (1994) *Ingenium und Individuum: eine sozialwissenschaftliche Theorie der Technik und Wissenschaft*, Springer Verlag, Wien

Schmutzer, M. and Bandler, W. (1980) 'Hi and Low – In and Out: Approaches to Social Status', *Journal of Cybernetics*, vol 10, pp283–299

Stone, D. (1988) *Policy Paradox and Political Reason*, Scott Foresman, Glenview, IL

Stone, D. (1997) *Policy Paradox: The Art of Political Decision Making*, W. W. Norton & Company, London

Thompson, M. (1996) 'Inherent Relationality: an Anti-Dualist Approach to Institutions', *LOS-Centre Report 9608*, Bergen

Thompson, M. (1997) 'Security and Solidarity', *The Geogaphical Journal*, vol 163, no 2, pp141–149

Thompson, M. and Ney, S. (1999), 'Consulting the Frogs: the Normative Implications of Cultural Theory', in M. Thompson, G. Grenstad and P. Selle (eds) *Cultural Theory as Political Science*, Routledge, London

Thompson, M and Rayner, S. (1998) 'Cultural Discourses' in S. Rayner and E. Malone (eds) *Human Choice and Climate Change, Volume 1: The Societal Framework*, Batelle Press, Columbus, OH

Thompson, M. and Schwarz, M. (1990) *Divided We Stand: Redefining Politics, Technology and Social Choice*, Harvester Wheatsheaf, Hemel Hempstead

Thompson, M. and P. Tayler (1985) *The Surprise Game: an Exploration in Constrained Relativism*, Institute for Management Research and Development, University of Warwick, Leamington Spa

Thompson, M., Wildavsky, A. and Ellis, R. (1990) *Cultural Theory*, Westview Press, Boulder, CO

Thompson, M., Grenstad, G. and Selle, P. (1999) *Cultural Theory as Political Science*, Routledge, London

TÜV   (2008),   'TÜV',   http://en.wikipedia.org/wiki/Technischer_%C3%9 Cberwachungsverein, accessed 16 October, 2008

Weinberg, A. (1972) 'Science and Trans-Science', *Minerva*, vol 10, pp 209–222

Wildavsky, A. (1987) *Speaking Truth to Power: the Art and Craft of Policy Analysis (2nd Edition)*, Transaction Press, New Brunswick, NJ

Williams, O. E. (1975) *Markets and Hierarchies: Analysis and Antitrust Implications: A Study in the Economics of Internal Organization*, Free Press, New York, NY

3

Transport

Introduction

Being on the move seems to have become part of our lives today. The distances between places we need to be – our homes, our workplaces or our families – have grown in line with our abilities to get about. Take me for example. As I am writing these lines, I am sitting in an office in Singapore. My parents live in Germany and my in-laws live in Austria. If my two little children want to visit their Grandma, or Oma, they need to travel halfway across the globe. When I was growing up, my Oma lived on the other side of town. Now, while our situation may not be entirely representative, it does point to a general trend of the past decades: our societies have become mobile.

It is probably not surprising, then, that issues surrounding how best to promote, organize and regulate this mobility is one of the most contentious questions on contemporary policy agendas. The reason, as we shall see shortly, is because modern transport systems shape our societies in very significant ways. Transport infrastructure creates the physical environments in which we live (try imagining your home town without roads). Our complex economies depend on transport systems to get raw materials to producers and finished goods to consumers. And without reliable and fast means of getting about, people – not just my children – would no longer be able to stay in touch with friends and family. In many ways, then, organizing mobility implies organizing society.

This chapter analyses the scope, structure and impact of conflict about transport policy in Europe. Before dissecting this intractable conflict, the next sections outline what makes the transport issue so messy. The following sections of the chapter apply the analytical framework outlined in the previous chapter to make sense of the transport policy debate. This section compares three contending policy stories – based on the policy frames outlined in the previous chapter – about transport policy in Europe. These policy stories help people dealing with transport issues to make sense of the complexities and uncertainties surrounding many of the key variables. Then the chapter goes on to discuss the way these policy stories create a landscape of agreement, disagreement and mutual rejection. The chapter ends with a discussion of the potential impacts of each policy story.

The transport policy issue

Transport policy, once a symbol for policy-making prowess, has become embroiled in divisive conflict. Developments that policy-makers, researchers and the public alike welcomed as firm evidence of social progress not even two decades ago are now disputed and contentious. Yet, unlike other messy policy challenges – such as global climate change – transport problems are, it would seem, fairly evident (as anyone who lives in a European or North American city will readily agree). Indeed, contemporary transport policy literature, be it an EU policy document, a critical scholarly analysis or a populist call to arms, points to the following four issues: the rapid growth in transport demand, the modal shift that satisfies this demand, the changing patterns of land use, and the increasing costs of these developments.

Increasing European demand for transport

Transport demand in Europe has grown rapidly since World War II. 'Our welfare societies', contend Himanen et al (1993), 'have apparently generated a complex array of contact patterns (material and immaterial) which require physical interaction at an unprecedented scale' (p8). The spatial development of Europe, they tell us, has been a 'geography of movement' (Himanen et al, 1993, p2): demand for moving of goods and people is at an unmatched level and is set to rise in the future.

Passenger transport in Europe has continuously grown in the past decades. Between 1995 and 2004, it grew in the European Union (EU-25) by 18 per cent at an average growth of 1.8 per cent per year (EUROSTAT, 2007). In the EEA (EU-25 + Iceland, Liechtenstein and Norway), passenger transport increased by a staggering 30 per cent between 1990 and 2002 (EEA, 2006). Between now and 2020, the European Commission (2006) expects passenger transport to grow by another 35 per cent (p26).

Freight transport has increased even faster. Over the past three decades or so, demand for the movement of goods has increased from 700,000 million tonne-kilometres in 1981 to ca 1,100,000 million tonne-kilometres in 1994 (European Council of Ministers of Transport, 1998). Since then, freight transport has grown at an average rate of 2.8 per cent in the EU-25 countries (EUROSTAT, 2007). By 2020, freight transport is expected to have increased by 50 per cent (European Commission, 2006, p26).

In both cases, the growth in transport demand mirrors the development of GDP. In the period 1995–2005, the 31 per cent increase in goods transport in the EU-25 countries outstripped economic growth at only 25 per cent (EUROSTAT, 2007). At 18 per cent, passenger transport in the EU-25 countries has grow a little slower than GDP in 1995–2004 (EUROSTAT, 2007). However, in 1990–2002, passenger transport in the EEA countries outgrew the development of GDP (EEA, 2006). In 2004, the transport sector employed 8.2 million people in Europe and added EUR 363 billion to European GDP (EUROSTAT, 2007). It would seem, then, as if freight and, to a lesser extent, passenger transport, drive economic growth in Europe.

Meeting demand on the road

The expansion of road transport has absorbed most of the spectacular rise in transport demand (Himanen et al, 1993; Banister and Button, 1993; European Commission, 1995; European Commission, 2001, 2006; EURO-STAT, 2007). Owing to the flexibility and independence that private cars afford, they are responsible for about three-quarters of all passenger kilometres travelled in the European Union (EUROSTAT, 2007). Including buses and coaches, about 85 per cent of all passenger transport took place on the road in 2004 (EUROSTAT, 2007). Air transport, which has shown the most rapid expansion since the mid 1990s (EEA, 2006), accounted for about 8 per cent, while the railways trailed at 5.8 per cent of passenger-kilometres travelled in Europe (EUROSTAT, 2007, p5).

Freight transport presents a similar picture. The changing nature of logistics and stock management (e.g. just-in-time distribution) has given road haulage a comparative advantage over other transport modes (Banister and Button, 1993, p3; CLECAT, 2005). Excluding sea transport, the EEA (2006) reports the share of road haulage for the EU-25 countries in 2003 to be 76 per cent (p40). Even if one includes sea transport within the EU, the share of freight on the road still dominates at 44.2 per cent compared to 39.1 per cent transported at sea (EUROSTAT, 2007, p5). According to EUROSTAT (2007), the period 1995–2005 saw road transport grow most (37.9 per cent) followed by sea (34.6 per cent) and, notably, air transport (31.1 per cent) (p69). Rail, however, only grew by 9.2 per cent in the same period and only made up 10 per cent of freight transport in 2005 (EUROSTAT, 2007, p5).

Road transport is likely to remain dominant in the European modal mix for some time to come. Transport projections suggest that, in 2030, more than 75 per cent of passenger-kilometres and something like 77.4 per cent of freight will move on the road (European Commission, 2003).

For planning purposes, the European Commission (2006) assumes that, by 2020, road haulage will have grown by 55 per cent while private car use will increase by 35 per cent (with air travel more than doubling) (European Commission, 2006, p26).

Changing patterns of land-use

Another distinct feature of the European transport policy issue is the changes in land-use patterns that have both caused and exacerbated present transport trends. All major European cities, maintains Peter Hall (1995), have decentralized since World War II. Banister and Button (1993) argue that the prices for inner-city housing, and an increase in income levels that seems to kindle the desire to own a car, as well as people's yearning for space, have depopulated urban areas (Banister and Button, 1993). As a result, complex and longer car journeys have replaced the 'simple journey-to-work pattern' (Banister and Button, 1993, p2). For this reason, EU citizens travelled an average of 32km per day (excluding air and sea-travel). Of these, car travel made up 27km (EUROSTAT, 2007, p112). Invariably, this has led to more traffic on the road. The bottom line, Hall maintains, has been a significant increase in journey lengths with no obvious savings in journey times despite ever-increasing capacities for speed in all transport modes (Hall, 1995).

The costs of transport growth

For a long time policy-makers assumed the benefits of modern transport to outweigh the costs. However, the size and complexity of contemporary transport systems incur new types of costs in the form of congestion, pollution and accidents. Taken together, many argue, these costs make modern transport systems seem far less of a good bargain.

The combination of rapid growth in transport demand, the reliance on road transport, and the changing patterns of land-use have generated significant congestion on roads across Europe (European Commission, 1993, 2001, 2006). European cities are close to their absolute limits, and observers point to the increasing ineffectiveness of new road investment in reducing congestion (Banister and Button, 1993; Himanen et al, 1993; European Commission, 1995). In 1993, the European Commission estimated the costs of congestion to be in the region of 2 per cent of European Union gross domestic product. In 2001, the European Commission (2001) projected road congestion in Europe to increase by 142 per cent at a cost of €80 billion – which amounts to 1 per cent of Community GDP – per year (European Commission, 2001, p8).

The pollution caused by transport represents another cost. Modern transport systems cause environmental problems at local, regional and global levels (Banister and Button, 1993). At the local level, the noise from vehicles – be they cars, heavy goods vehicles, trains or aeroplanes – has matured from being a mere nuisance to an environmental problem of significant proportions. At the transboundary and global levels, emissions from vehicles contribute to acid rain, stratospheric ozone depletion, concentration of tropospheric ozone and global climate change (Banister and Button, 1993, p4). Currently, the transport system consumes almost 83 per cent of all energy and accounts for 21 per cent of GHG emissions in the EU-15 countries (EEA, 2006; EUROSTAT, 2007). Since 1990, emissions have steadily increased and are set to grow by something like 10 per cent between 2005 and 2015 (EEA, 2006). In terms of human health, the European Commission (1995) estimates that '... air pollution from transport kills more than 6000 people in the UK alone' (European Commission, 1995, p1). It all adds up: in 2006, the European Commission placed environmental costs at about 1.1 per cent of European GDP (European Commission, 2006, p8).

Last, European accident statistics show that increased transport demand, coupled with increasing vehicle capacities, has come at a grisly price. In 2005, 43,000 people were killed in transport accidents, 41,300 of these on the road. In 1995, the estimated financial costs of transport accidents stood at about 1.5 per cent of European Union GDP (European Commission, 1995, p22).

The transport dilemma

It would seem, then, as if transport policy-makers are on the horns of a dilemma. On the one hand, high levels of mobility not only drive economic growth but have become defining features of our prosperous life-styles. On the other hand, mobility causes increasingly high costs in terms of the environment, human health and social balanced mobility. Do nothing, and congestion will slow economic growth and undermine our way of life. Build more roads, and environmental degradation as well as social pathologies will undercut any benefits of economic growth.

How, then, are transport policy-makers to resolve this dilemma?

The scope of policy conflict in European transport policy-making

Analysis of the European transport policy debate, using the framework developed in Chapter 2, reveals three competing policy narratives. The following analysis breaks down each narrative into its component parts: each story creates a particular setting (the assumptions), identifies villains (the problems) and provides heroes to save the day (the solutions). In this way, each narrative links cause and effect, identifies specific culprits and suggests a particular course of action. Significantly, each policy story resolves the transport policy dilemma in a different way.

Efficient mobility

The first story tells a tale of how modern transport systems drive economic growth and prosperity. The tale, told by industry actors and their supporters in politics and research, likens modern transport systems to arteries, and the mobility they provide to the lifeblood of dynamic, innovative and competitive societies. Without modern transport systems, so the argument goes, our prosperous life-styles are unthinkable. Policy, then, needs to unravel the transport dilemma by promoting innovative and competitive solutions to transport problems; this, in turn, means ensuring that transport systems meet the growing demand for mobility on which our well-being and innovative capacity depend.

The setting: the mobility–opportunity–prosperity cycle

In a world of opportunities, mobility drives growth and prosperity. Each of us, Ted Balaker (2006) of the Reason Institute argues, stands in a 'circle of opportunity': the larger the circle, the more opportunities we have for creating wealth. As our mobility increases, so does the circle opportunity.

Once we turn these opportunities into wealth, we can afford faster and safer means of getting about. The further and faster we move, in turn, the larger our circles of opportunity become. Thus, mobility creates wealth which in turn increases mobility which creates more wealth and on and on it goes (see Box 3.1). On this view, mobility, opportunities and prosperity are linked in a virtuous and mutually reinforcing cycle.

In market economies this virtuous cycle propels economic growth (CDU, 2007; Balaker, 2006). There is, the European Automobile Manufacturers' Association (ACEA) tells us,

> ... *a correlation between economic growth, population growth and transport need. The correlation factor may vary from time to time but without efficient transport Europe would not have the position in the world economy it has today. (ACEA, 2002)*

This is because economic growth is necessary for anything that societies may want to do. Protecting the environment, showing solidarity with the developing world or building an inclusive society require resources. Mobility, by driving economic growth, helps us obtain these resources. It follows, proponents of the efficient mobility story contend, that whatever our policy goals may be, reaching them requires the mobility–opportunity–prosperity cycle to do its magic. All that transport policy-makers need do is to ensure that the virtuous cycle spins freely and vigorously.

In practice, so proponents of the efficient mobility story argue, this means doing two things. First, transport policy-makers need to provide efficient transport infrastructure. Since cars, buses and HGVs 'offer the most flexible, convenient, time-efficient mode of travel and transport' (ACEA, 2002, p4), much of this infrastructure will necessarily be roads. Second, for increased mobility to translate into economic growth, entrepreneurs need the freedom to compete. This is why European car manufacturers believe '... in a free market. The motor industry welcomes competition – fair competition – and believes that customers should also be allowed to choose what forms of transport they wish to use' (ACEA, 2002, p5). This also means allowing the free market to weed out uncompetitive firms and inefficient ways of getting about (CLECAT, 2005).

Box 3.1 *Classic transport planning*

For a long time, transport planners believed that the ease with which individuals could reach locations determined economic growth. This positive relationship between mobility, accessibility and wealth, Banister and Lichfield (1995) tell us, is at the heart of classical location theory. Here, the degree of accessibility determines the monetary or utility values of different locations. Transport costs, this approach argues, are a key component of land and rental values. Price and rent levels, in turn, determine land use. As transport costs have fallen and people have become less averse to longer journeys, so the story goes, suburban communities have become increasingly attractive (Banister and Lichfield, 1995, p7).

The underlying rationale of this approach was developed in the USA back in the 1950s. At this time, research started to look carefully at the relationship between spatial organization and location. Analysis of transport patterns in US cities revealed that accessible locations attracted investment while inaccessible locations did not. This suggested that the growth of transport and the development land-use patterns in some way reinforced each other. On this view, specific land-use patterns emerged because of the location of activities such as work, shopping or leisure. Getting to and from these locations required transport. The development of transport infrastructure to service these locations increased their attractiveness for investment. The resultant growth of economic activity would not only entrench the specific pattern of land use but would also increase the demand for transport and so on and so forth (Wegener, 1995, p175). Insights such as these led early transport planners to conclude that:

- transport shapes cities;
- transport policy affects the spatial organization and development of cities;
- transport is a function of land use;
- land-use policy influences transport (Wegener, 1995, p159).

This 'land-use–transport feedback cycle' implied an important role for rational transport planning. Since journey and location decisions were co-determinate, American planners concluded that they had to consider transport and land-use planning together (Wegener, 1995, p157). If transport determines the economic development of locations, then the rational planning of transport can bring about land-use patterns that drive economic growth.

The villains – congestion

For the proponents of the efficient mobility story, anything that limits our mobility reduces opportunities, lowers economic growth thereby leaving us with less resources to fund our cherished policy projects. Today, so the efficient mobility story argues, congestion slows us down more than anything else (Balaker, 2006; ACEA, 2007; CLECAT, 2005; Banks et al, 2007). When you are stuck in traffic, opportunities go to waste. For each of us, congestion imposes costs in the form of '... stress, lost time, lost money from extra gas and wear and tear on automobiles' (Balaker, 2006, p3). Aggregated over the entire economy, congestion erodes economic growth. For this reason, the RAC argues that ...

> ... [u]nless congestion is tackled, not only will journeys on the roads become even slower and less reliable than they are today, the country's international competitiveness, and hence economic growth, will be damaged. Businesses will look to locate abroad for better connections to markets, suppliers and workforces. Quality of life will suffer as the increasingly diverse variety of journeys which are an important part of modern life become more difficult and time-consuming. (Banks et al, 2007, p8)

Congestion, so proponents contend, has resulted from misguided transport policy. Basically, so the argument goes, policy-makers have not been doing their jobs. For one, policy-makers across Europe have failed to expand transport infrastructure to meet burgeoning demand (Balaker, 2006; European Commission, 1993, 2001). ACEA (2002) laments that for ...

> *... years many billions of euros have been raised by taxation, duties, tolls, fines and other charges which have been used by member states' treasuries to finance almost everything other than good infrastructure projects or good driving education and training. (p4)*

Despite projected increases in transport demand, under- and mis-investment in transport capacities, particularly in roads, has left existing transport infrastructure stretched far beyond its limits (Banks et al, 2007; Balaker, 2006). Clogged roads across the continent have been the inevitable result.

Rather than concentrating on what they should be doing, policy-makers have instead meddled with transport markets. For example, large chunks of tax-payers' money have gone towards protecting public transport from competition (CLECAT, 2005; ACEA, 2002, 2007). Worried about the low and falling share of the railways in passenger and freight transport, policy-makers have erected a veritable fortress of regulation and subsidies around railways. But, as the European Association for Forwarding, Transport, Logistics and Customs Service (CLECAT) (2005) pointedly observes, the ...

> *... problem is not so much the road but the failure of rail freight to offer the same level of efficiency as the road. This unwanted result is due to the fact that whilst everyone was very busy building a common market since 1993, the rail market has been very busy preserving borders and divisions that de facto hamper competition. (p3)*

Predictably, protection has not worked. After decades of unfair subsidies and taxes ostensibly aimed at 'internalizing the external costs of road transport', public transport is simply not up to the job (ACEA, 2002; CLECAT, 2005). Instead of innovating and meeting customer needs, misguided protectionism has allowed the railways to languish in a state of inefficiency (ACEA, 2002, 2007; CLECAT, 2005). In contrast, the discipline of the market has forced road transport in Europe to innovate and compete. As a result, rail has become too inflexible for passengers and too expensive for haulage. Unless exposed to real competition, proponents argue, rail is unlikely to improve on its 10 per cent share of transport: firms, so the argument goes, '... are not going to choose rail freight just because they feel it is more environmentally friendly. They will only choose rail solutions if they are either as efficient or more efficient than road solutions' (CLECAT 2005, p2).

How could transport policy-makers have taken such a wrong turn? The reasons, advocates of mobility contend, are to be found in poor and politically motivated analysis. Transport policy-makers in Europe rely on shaky concepts, such as 'social marginal cost pricing' or 'transport externalities', to design measures and programmes. As theoretically elegant and ideologically appealing as these concepts may be, there is very little evidence that they will work where, quite literally, the rubber meets the road (see Box 3.2) (ACEA, 2002, 2006; CLECAT, 2005).

Box 3.2 *External costs as junk science*

Market actors in the European transport are unconvinced of the scientific arguments that underpin much of current transport policy. Transport policy debate in Europe, so ACEA (2007) argues, has been based on the fact that an ...

> ... *eclectic and sometimes arbitrary mix of statistical sources is used in the urban transport debate. Some of these lack authentication and their use adds little to the credibility of the discussions. Some of them may give a misleading picture. Some of the existing sources give data from only a small number of Member States, which may not be representative of the EU as a whole or should not be extrapolated to the entire Union. (p11)*

On the one hand, proponents of the efficient mobility story argue, the figures on which the European Commission bases its argument are suspect. For example, ACEA (2001) questions whether road transport contributes as much to CO_2 emissions as usually claimed. Instead of being the fastest growing source of CO_2 emissions, some studies suggest that '... CO_2 production from road transport would stabilize in 2005 despite the increase in traffic' (ACEA, 2001, p15).

On the other hand, even if these claims were to be true, proponents of the efficient mobility story doubt that marginal social cost pricing is an appropriate tool. ACEA (2001) contends that there is not a shred of evidence to support the more enthusiastic claims about the efficacy of internalizing external costs of transport. Moreover, the European Commission would need an accurate evaluation of external costs for these pricing mechanisms to work: there is, however, '... no prospect of a robust calculation of this kind being available for many years' (ACEA, 2001, p8).

Ultimately, advocates of mobility conclude, it all comes down to politics. Since democracies do not reward policy-makers for taking tough and unpopular decisions, the issue of congestion remains largely unresolved: transport planners in the USA, so Balaker (2006) contends, '... put off dealing with the problem for another year, then another, and another. Along the way, the American people have coped with degraded mobility by changing their behavior' (p11).

The heroes – let the cycle spin

If congestion is slowing us down, so champions of the efficient mobility story contend, policy needs to find ways of bringing us back up to speed. To do this, governments just need to concentrate on what they do best (providing and managing infrastructure without interfering in the market) to allow innovative entrepreneurs to get on with what they do best (innovating, exploiting opportunities and creating wealth). Policy-makers ought not '... lose sight of the real requirement of a transport policy: to provide mobility for the citizens and products' (ACEA, 2002, p3).

What we really need, proponents of the efficient mobility story tell us, is transport infrastructure to absorb growing transport demand. CLECAT (2005) believes that the focus of transport policy ...

> ... *should be on the rapid development of infrastructure. Improved infrastructure facilitates an improved and more efficient supply chain, which in turn contributes to greater competitiveness.* *(CLECAT, 2005, p5)*

Since, as the German CDU (2007) remarks, 'we will not be able to cope with tomorrow's traffic flows using yesterday's transport infrastructure' (p7), policy-makers will need to extend all transportation modes (ACEA, 2002).

Yet, instead of interfering with the market by favouring one mode over another, '[e]ach transport mode should be used where it is at its best and the choice of mode should be decided by market forces' (ACEA, 2002, p12). Letting the market decide, so proponents of the efficient mobility story tell us, means that the flexibility and efficiency of road transport will continue to dominate transport markets. Like it or not,

> ... *road transport is a necessity and will be the number one goods mode in the coming years ... Road transport can improve substantially by actions in the fields of road infrastructure (eliminating bottlenecks and improving road safety), weights and dimensions and the use of information technology (in logistics as well as in congestion avoidance and road safety). (ACEA, 2002, p11)*

Getting our societies moving, so the argument goes, calls for the construction of new roads, the extension of existing road infrastructure and more effective management of the road network. For example, the CDU (2007) points out that in Germany six-lane motorways must become the standard while motorways in congested areas should be extended to eight lanes (CDU, 2007). Similarly, Banks et al (2007) from the RAC Foundation argue that '... there is a strong economic case for more strategic road building in Great Britain at an annual rate of around 600 lane kilometres a year (Lkmpa), or more ...' (p11)

Efficient means of managing transport networks will also bring us up to speed. In urban areas, ACEA (2007) argues, policy needs to focus ...

> ... *not on penalization and reduction of traffic but on its fluidity to respond better to mobility needs. This can effectively be done through urban traffic planning to define transport policies for goods, passengers and infrastructure. (ACEA, 2007, p4)*

Advocates of mobility point to a wide range of technological and organizational methods for managing increasing traffic flows. These measures include the synchronization of traffic lights, the effective coordination of road works, multiple function lanes for different uses at different times of day, or implementing park-and-ride schemes (ACEA, 2007).

But what about the environmental and social costs? There can be no denying that contemporary transport systems can be a source of pollution, social dislocation and accidents (ACEA, 2002, 2007). However, mobility itself solves these problems through the innovation associated with growing prosperity. For example, economic growth funds research into fuel-efficient vehicles. In practice, so ACEA (2002) argues, this has meant that ...

> ... *vehicles have become progressively cleaner and cleaner. Passenger vehicles will produce 95 per cent fewer emissions than in the mid-70s. The latest emission limits in 2005 and 2008 will make further massive reductions to air pollution from road transport. Trucks will produce 90 per cent fewer emissions than in 1985. (p14)*

Given this trajectory, proponents argue that the automobile will '... cease to be a polluter in the conventional sense' (ACEA, 2002, p4). Indeed, advocates of mobility point out that consumer-driven technological advances have done far more to cut noxious emissions than any road charge or rail subsidy (ACEA, 2002).

Technological innovation not only improves vehicles, it also makes transport infrastructure more efficient and reliable. ITS (Intelligent Transportation Systems) provides road users with '... increased safety, better information, greater comfort and reduced journey times' (ACEA, 2007, p7). Designing and constructing 'intelligent' transport infrastructure, so the *Association Européeanne des Concessionnaires d'Autoroute et d'Ouvrages à Péage* (ASECAP) (2002) argues,

> ... *will lead to significant fallout in terms of traffic management, safety, information and services to users, emission reductions etc. and thus of sustainable development. The aim, undoubtedly, is to better evaluate possible gains and to ensure that the level of 'intelligence' of the vehicles corresponds to that of infrastructure equipment. (p12)*

All this, however, assumes that the mobility–opportunity–prosperity cycle is well greased. For this to happen, transport policy-makers – particularly those in Europe – need to abandon their deluded aspirations of steering transport flows and micro-managing modal splits (ACEA, 2002; CLECAT, 2005; CDU, 2007). If transport policy is to relieve congestion, then '[m]arket forces and practical circumstances should decide the modal split' (ACEA, 2002, p12). This means cutting back public sector transport regulation according to the principle of 'as much market as possible, as much state steering as necessary' (CDU, 2007, p7). Apart from financing transport infrastructure, then, proponents of the efficient mobility story see very little for the public sector to do except to '... ensure that all service providers enjoy a level playing field by abating dominant positions and strongly resisting any attempts to restore them' (CLECAT, 2005, p6).

Sustainable mobility

The sustainable mobility story tells of the complicity between modern transport systems and pervasive inequities, exploitation and suffering. Unfair transport and land-use policies have left us with environments incapable of fulfilling our most elementary social and biological needs. Congestion, accidents, commuting and, particularly, all kinds of pollution are depriving people of their basic human rights to a liveable and healthy life.

This is an angry story driven by outrage and disgust. Proponents of this story, mostly environmental NGOs (such as Transport and Environment, or Friends of the Earth) and political parties committed to environmental issues, i.e. national and European Green parties, urge us to resolve the transport dilemma by abandoning a grossly unjust and inequitable socio-economic order.

The setting: holistic transport policy

Modern transport systems, so proponents of the sustainable mobility story argue, are fundamentally unfair. Just like the exploitative and unjust socio-economic order in which they are embedded, transport systems distribute costs and benefits of mobility inequitably. They are systems in which the vulnerable members of society – women, children, older people, ethnic minorities or the poor – find it difficult and dangerous to get around. Modern transport systems destroy our natural and social environments to serve the wants of a privileged few. They are systems that revolve around road transport because powerful economic interests, such as the automobile and construction industry, bend political processes to their selfish will and against the well-being of most citizens (Knoflacher, 1997; T&E, 2003).

Why, then, have citizens not seen through this swindle? The reason, proponents maintain, is because we all are trapped in a view of the world that draws artificial distinctions between people and things. The theology of profit and its doctrine of greed obscures our view of the intrinsic connections between the natural and social worlds. It also prevents us from understanding that all humans have the same basic needs. On this view, we all occupy the same vulnerable social and ecological space.

For proponents of the sustainable mobility story, the holistic view of society and nature has two important consequences. First, realistic assessment of transport systems, so the argument goes, sees transport as an inextricable part of the larger natural and social system. That is why, the German Green Party argues, '... measures planned in the transport sector must be set against criteria of ecological, social and economic compatibility' (Bündnis90/Die Grünen, 2002, p28). Second, proponents of the sustainable mobility story argue that a holistic approach allows us to keep sight of the really important things in life. Rather than craving ever more consumer goods, a holistic view affords a clear view of our real human needs: the need for healthy (meaning equitable and just) social and ecological environments.

Most importantly, a holistic approach suggests that transport policy alone cannot solve contemporary transport problems. Significantly, many causes of transport policy problems are rooted in fundamental socio-economic, political and ideological practices. Sustainable transport policy means tackling these rudimentary societal practices.

The villains: unsustainable and inequitable transport systems

Today's transport systems, proponents of the sustainable mobility story argue, are killing us. The patterns of contemporary mobility – something John Adams (2005) calls 'hypermobility' – are destroying our natural and social habitats.

Pollution, particularly from road and air transport, has degraded air quality throughout Europe (EEA, 2006). Unabated growth in CO_2 emissions from transport are driving the world towards the 2 degrees threshold of 'dangerous climate change' (EEA, 2006). Noise pollution means that a decent night's sleep has become a rare luxury for most Europeans, particularly city-dwellers (Bündnis 90/Die Grünen, 2002).

Hypermobility also corrodes social environments. Road traffic has made the local neighbourhood – traditionally a place where children acquired crucial interpersonal and social skills – inhospitable and dangerous environments. As a result, unsupervised journeys by children have substantially decreased: while not even 20 years ago, British children were regularly walking or cycling to school, to sports activities or to friends, they have now become '... captives of the family chauffeur' (Adams, 2005, p4). Similarly, Knoflacher (1997) shows how high-speed transport links between central and peripheral regions drain the life-energy of the economically weaker locality. As work and services, such as shopping, health care and government services shift from the periphery to the centre, the indigenous transport patterns that sustain the fragile socio-cultural network are irretrievably destroyed (Knoflacher, 1997, pp57–58).

By degrading our habitats, modern transport patterns extort a grim price in terms of physical and mental health. Over the past decades, the incidence of respiratory diseases due to transport-related air pollution has grown alarmingly. Increasingly sedentary life-styles have led to an increase in obesity and obesity-related health problems, particularly in those children who cannot go outside to play (Hillman, 1999). Across Europe, stress-related mental health problems are on the rise (Bündnis 90/Die Grünen, 2002). This is not counting the 80,000 deaths on the road across Europe every year. Against the backdrop of these figures, the German Green Party (2002) argues that although ...

> ... *freedom of movement and travel are an integral part of an open society, motorized vehicles impair life quality in cities, towns and the countryside: They causes [sic] noise, tailbacks, and an environmental damage that, at its extreme, leads to climate collapse; they makes people ill, and year after year causes unacceptably high numbers of fatalities and injuries. (p28)*

Worse still, so proponents of the sustainable mobility story continue, costs and benefits of transport are unequally distributed. While the rich enjoy all the benefits of fast, convenient and continent-spanning travel, the poor, stuck in over-crowded, dirty and dangerous places, are left to deal with the consequences (Adams, 2005). Dominance of road transport isolates many citizens from social, economic and cultural life. Not owning a car, perhaps because you cannot afford one, can mean that you find yourself cut off from many activities such as out-of-town shopping, sporting events, the workplace and hospitals and health care (T&E, 2003). In terms of road safety, people living in poorer parts of Europe, the ETSC (2002) reports, are exposed to considerably higher accident risks than populations in more affluent regions. In addition, the ETSC (2002) identifies ...

> ... *substantial differences between the safety of the different categories of users within the road mode. While car users comprise the greatest proportion of overall road deaths (57 per cent), the risk of death on EU roads is substantially higher for vulnerable road users – some 8–9 times higher for pedestrians and cyclists. (p1)*

Similarly, impaired social and spatial mobility of the poor and vulnerable makes it difficult for them to avoid air and noise pollution. As climate change sets in, the poor in the developing world will have to deal with the impacts of transport-related pollution from rich countries (Knoflacher, 1997). Finally, even in rich countries, advocates of sustainability contend, we have burdened future generations, who have no voice today, with the costs of transport-related pollution. 'The price of today's economic growth', Knoflacher avers, 'will be paid by the workers, the social system, the 'Third World' and the next generation that will have to inherit a destroyed planet' (Knoflacher, 1997, p119).

Why, then, do we continue down this ruinous road?

Part of the reason is that many still believe in the myth of the mobility–opportunity–prosperity cycle. Conventional transport planners, Knoflacher (1997) argues, believe that transport infrastructure investments save time on journeys. A holistic approach, however, shows that transport is a zero-sum game: a new motorway may save motorists ten minutes but will add at least the same amount of time to cyclists' journeys, who now need to take a more circuitous route to avoid the motorway (Knoflacher, 1997, pp43–48). In the same vein, John Adams (1995, 2005) and Mayer Hillman (1999) argue that road safety measures such as safer cars, seatbelts or wider roads merely make motoring safer for motorists. The increased speeds that road safety measures engender, disproportionally increase risks to pedestrians and cyclists. The decline in road accidents merely document the fact that, since roads are now far too dangerous, non-motorists have all but retreated from the transport system. Yet, since journey times and transport safety are distributed inequitably, the illusion that modern transportation saves time lives and flourishes.

As a result, proponents contend, European transport policy tends to reflect narrow interests, such as economic growth, but ignores important aspects that determine the quality of life, such as road safety and the socio-cultural aspects of transport systems (T&E, 2003; ETSC, 2005). Anyone in doubt, argue proponents, is advised to look at the way transportation modes in Europe are taxed. Despite the wailing and breast-beating about the burden of taxation in the road sector, '... total taxes as recorded by the European Automobile Manufacturers' Association (ACEA), €360bn, only recoup about half of the total social costs (€682bn). Hence cost coverage is just over 50 per cent' (T&E, 2007). Similarly, prevalent subsidies and tax breaks in the European aviation industry '... are very environmentally harmful as they do not incentivize sustainable lifestyles and create artificial demand in a sector that is already very far from having internalized its external costs' (T&E, 2007, p11).

The myth of the mobility–opportunity–prosperity cycle and its accompanying analytical blindness not only afflicts policy-makers and planners. In developed and developing societies, the car has come to symbolize freedom and power. Car ownership is a measure of social value; the bigger and faster the car, the more we believe we are worth. People want to own cars, proponents of the sustainable mobility story argue, regardless of whether this is what they really need. Until these ingrained attitudes towards car ownership change, T&E (2003) concludes,

> ... *non car-owners will typically want to own one. It is therefore no surprise that people in the lowest income groups frequently make a car their first purchase when they have some money, or that buying a 'better' car is an important symbol of success. (p2)*

Another part of the reason why we are belting down the fast lane towards an ecological and societal pile-up, proponents contend, is that hypermobility suits some people very nicely. Here, Knoflacher points to an unholy alliance of politics, big business and research. For years, the researchers in this iron triangle have provided the conceptual tools to hide the special interests behind transport policy: those of the motor vehicle industry and the construction industry. This is why proposals for necessary change in European transport policy meet with lukewarm responses on the part of policy-makers. For example, the ETSC (2008) notes that European governments are dragging their feet when it comes to the implementation of road safety measures. Logic, let alone morality, so proponents argue, is turned on its head to suit big business interests (Knoflacher, 1997).

The heroes: slowing down

If we are to avoid an ecological catastrophe, proponents of the sustainable mobility story argue, we will need a radically different transport system. We need to change the very nature of mobility and its place in the socio-economic system. The German Green Party envisages such a transformation as follows:

> *Our goal is an ecological circular-flow economy where resources and energy are used economically and efficiently. Regional circulation of goods has to take priority over long-distance transportation chains, and foodstuffs are to be produced locally, where possible. This is a trend that will be encouraged by consumer demand. Environmental compatibility must be a central criteria for research and development. In order to re-establish the global ecological balance and secure the necessities of life for a growing world population, we need an ecological-technical revolution that will reduce the consumption of resources by a factor of ten within the next decades. (Bündnis 90/Die Grünen, 2002, p21)*

Bringing this vision about entails a three-pronged policy strategy. First, sustainability requires a radical reduction of demand for transport. The best, but not necessarily easiest, way to reduce transport demand, so the argument goes, is to obviate the need for mobility by adopting new principles of land-use, urban and transport planning. In practice, the German Greens argue, this means:

> *... bringing housing, work and leisure closer together, cutting traffic instead of encouraging it, protecting people and conserving landscapes rather than encouraging further land use, keeping and modernizing existing networks, putting noise protection measures above simply building anew, and improving systems rather than expanding them. (Bündnis 90/Die Grünen, 2002, p28)*

Recalibrating land use and urban planning in this way will 'shrink' transport networks. Smaller networks covering shorter distances, in turn, replace rapacious 'hyper-mobility' with sustainable micro-mobility (see Box 3.3).

Box 3.3 *Micro and macro mobility*

Not all mobility, proponents of the sustainable mobility story contend, is necessarily bad. Indeed, certain types of mobility are vital for the vibrancy and health of small, self-contained and autonomous communities egalitarians prefer. This is what Hermann Knoflacher (1997) calls 'micro-mobility': slow, life-sustaining and environmentally sustainable mobility within small regional areas. However, exploitative, inequitable and unsustainable modern transport systems promote what Knoflacher calls 'macro-mobility': the high speed, socially and environmentally damaging movement of people and goods across vast distances. Similarly, John Adams (2005) refers to these mobility patterns as 'hypermobility'. Whereas policy should be promoting micro-mobility and thus reducing the speed of the transport system, policy-makers constantly seek to increase speeds thus destroying humans' social and environmental habitats.

Knoflacher continues that the rising demand for 'macro-mobility' is not a sign of social vitality. On the contrary, our constant urge to move points to a sense of deep discomfort. For Knoflacher, the demand for mobility is a sign of unsatisfied need. Our frustration with a socio-economic order that cannot and will not meet our human needs makes us as fidgety as a child with ADHD. It is no surprise that people who are cooped up in poorly planned, dirty, dangerous and unpleasant cities and towns feel the desire to leave at the weekend. What is more, the increasing trend of individual atomization caused by advanced capitalist labour markets exacerbates the desire for mobility: a person who has a functioning social network and healthy interpersonal relations does not need to travel. The demand for increased mobility is a result of social pathologies such as single-parent families and single households.

Second, policy needs to encourage the use of sustainable transport. This means curbing the use of cars and aeroplanes (Bündnis 90/Die Grünen, 2002; T&E, 2003, 2007). A simple way of doing this, proponents argue, is to use taxation and road pricing to make the price of road and air transport reflect true social and environmental costs (Bündnis 90/Die Grünen, 2002; T&E, 2003, 2007). Advocates of sustainability also plan to use more direct command-and-control measures, such as speed limitations or driving bans (Bündnis 90/Die Grünen, 2002). In place of cars and aeroplanes, proponents of the sustainable mobility story envisage innovative forms of public transport. For example, car sharing, taxi schemes or combined public transport plans reduce the dependency on individual cars (Bündnis 90/Die Grünen, 2002). This requires, proponents contend, massive investment in the renewal of public transport in order to blur the lines between individual and public transport: 'Public transport services', the German Green Party tells us, 'must become more individual and attractive; individual transport must become more public and socially-oriented' (Bündnis 90/Die Grünen, 2002, p28).

However, advocates of sustainability know that neither the internalization of transport costs nor investment in public transport alone can convince people to switch to more sustainable forms of transport. Sustainable transport policy must overcome deep-seated psychological and socio-cultural prejudices concerning modes of transport (Bündnis 90/Die Grünen, 2002; T&E, 2003). Sustainable mobility, proponents contend, will have made great progress once the car ...

> ... *is no longer seen as the obvious choice of transport, providing status and other non-transport benefits: it is just a car. And a plane is just a plane. People's perceptions determine the nature of the transport system: so sustainable transport systems help people to be aware of the social consequences of their transport actions and encourage people to take individual responsibility for them. (T&E, 2003, p1)*

Last, sustainable transport policy must ensure that the most vulnerable do not pay for the unnecessary mobility of the rich and powerful. Citizens' environmental and health needs must be at the heart of transport policy-making. Policy-makers need to use the legislative and regulative means available to them to protect citizens from the adverse effects of modern transport systems. For example, since people living near airports have as much of a right to a decent night's sleep as the rich in leafy suburbs, policy-makers need to ban night-time landings and take-offs (Bündnis 90/Die Grünen, 2002). The best way of ensuring that the needs of the poor and vulnerable are heard, proponents of this story argue, is to provide them with a voice in the transport policy-process. Here, the German Green Party argues that ...

> *[i]ntegrated traffic planning means letting those affected participate. We are committed to citizen involvement in planning processes and passenger consultative bodies where a common point of reference can be established, based on the different mobility needs of women and men, young and old, or people with disabilities. (Bündnis 90/Die Grünen, 2002, p30)*

Bringing about sustainable mobility, proponents admit, cannot avoid treading on toes. Sustainable mobility is all about righting wrongs in the transport systems. And what is wrong with transport systems is that a few can enjoy all the benefits while the rest of us are stuck with the costs. 'Ensuring socially sustainable transport', T&E (2003) pointedly remarks, 'will mean upsetting some people: although society as a whole will benefit, those presently over-benefiting will have to give up some privileges' (p4).

Balanced mobility

The third policy story tells a tale of how transport planning and management have become extraordinarily complex tasks. As with the efficient mobility story, this tale understands modern transport systems to be an indispensable motor of prosperity and wealth. Like the champions of the sustainable mobility story, proponents of balanced mobility clearly see the unacceptable environmental and social costs of contemporary transport systems. Unlike with either of the rival stories, however, advocates of balance do not see transport policy-making in terms of stark choices. Instead of irresponsibly resolving the transport dilemma in terms of one or the other forces doing the pulling, advocates of the balanced mobility story argue that transport policy must find and maintain a point of equipoise between economic growth and prosperity on the one hand and social and environmental integrity on the other.

The setting: finding equilibrium

Transport systems, proponents of the balanced mobility story argue, are as vital to our ways of life as they are inseparable from them. On this view, transport systems are the basis of our prosperity: 'Efficient mobility of goods and persons,' argues the European Commission (2006), 'is an essential component of the competitiveness of European industry and services' (p6). But beyond the purely economic, transport systems also fulfil important social functions. At individual level, the former British prime minister Tony Blair argues that good transport '... provides access to jobs, services and schools, gets goods to the shops and allows us to make the most of our free time' (Department of Transport, 2004, p5). At regional level, transport networks can help us stitch back together the fragmented European social and economic landscape by simply creating connections between different parts of the continent (European Transport Workers' Federation, 2004). Whether on a micro- or macro-scale, so proponents of the balanced mobility story contend, transport networks enhance social balanced mobility and stability.

But, so the argument goes, modern transport has some unpleasant side-effects. As we have seen, modern transport leaves a large ecological footprint. In addition, proponents of the balanced mobility discourse are acutely aware that modern transport systems distribute risks across regions and users (ETSC, 2008). To both of these known issues, advocates of balanced mobility add a third: the uneven regional development in Europe. Unfortunately, so proponents of the balanced mobility story argue, these side-effects will not go away on their own (Banister and Button, 1993; Banister and Lichfield, 1995; Banister, 2008, see Box 3.4).

Box 3.4 *Doubts about classical transport planning*

For David Banister, a British transport economist, the two fundamental assumptions of the land-use feedback cycle no longer hold. First, Banister tells us that travel is no longer a derived demand. Evidence for the so-called 'escape theory' suggests that people undertake travel as a leisure pursuit in its own right (Banister, 2008, p74). Second, classical transport planners assumed that the most rational optimization strategy in transport was to minimize travel time by deploying increasingly efficient means of transportation. However, growing environmental and safety suggest that the '... key policy objective now becomes that of reasonable travel time, rather than travel time minimization'. (Banister, 2008, p74). In other words, Banister seems to suggest that efficiency – the holy grail of classical transport planning – may not be an objective and intrinsic feature of transportation systems. Rather, efficiency, just like social responsibility and environmental sustainability, is a value that policy-makers need to balance with other competing values.

Transport systems, then, are subject to forces that pull in different directions. On this view, giving in to any one particular force invites all sorts of problems. Too much economic growth degrades the environment and leads to social exclusion (Department for Transport, 2004; European Commission, 2006). By the same token, overly protective environmental regimes stifle economic growth, strangle innovative risk-taking and thereby exacerbate social exclusion.

If our societies are to thrive, so the balanced mobility story goes, policy-makers must find an equilibrium between the economic, environmental and social impacts of transport. For this reason, the UK Department for Transport points out that the ...

> ... *ability to travel offers all of us very real benefits and extending mobility is important in building an inclusive society. The transport system helps to underpin the international competitiveness of the economy. But mobility comes at a cost, whether financial, social or*

environmental. We need to ensure that we can benefit from mobility and access while minimizing the impact on other people and the environment, now and in the future. (Department for Transport, 2004, p11)

Striking this balance between costs and benefits, proponents of balanced mobility argue, requires hands-on management and planning (European Commission, 2001, p3). On this view, management and optimization of the transport system is not a short-term task to be taken on lightly. Transport policy decisions today, the Department for Transport (2004) says, '... will have an impact for decades to come. It is essential that we take a long term view' (p11).

The villains: imperfect transport markets

European transport policy-makers, proponents of the balanced mobility story argue, need to solve complicated policy problems. On the one hand, demand for transport is set to increase as economies grow and people become more wealthy (Department for Transport, 2004). On the other hand, existing transport infrastructure, already stretched to breaking point, is unable to absorb this growth.

Part of the problem, then, is persistent under-investment in transport infrastructures. In Europe, proponents point out, demand for transport has consistently outstripped investment in capacities of any transport mode (European Commission, 1993; European Commission, 2001; Department for Transport, 2004; SPD, 2007). Not only have national governments in Europe failed to provide adequate transport capacity, the European Commission itself has failed to complete the Trans-European Networks that aim to mesh transport systems of old and new Member States. 'The fact that not enough attention has been paid to developing infrastructures,' the European Commission noted as early as 1993, 'is one of the reasons for the deterioration in the quality of life' (European Commission, 1993, p22).

But, so proponents of the balanced mobility story point out, there is an added complication that rules out simply expanding transport infrastructure. Not only is capacity too low, transport networks are also severely imbalanced. As we have seen, across Europe – and also in the USA – transport systems are overly reliant on road transport. Moreover, in Europe, regional transport imbalances exacerbate modal imbalances. As economic and social life drains from peripheral regions into European centres, growing congestion in urban hubs contrasts increasing isolation in the regions (European Commission, 2001). The European Commission (2001), describes the 'paradox' as follows:

> *Saturation on some major routes is partly the result of delays in com-*
> *pleting trans-European network infrastructure. On the other hand,*
> *in outlying areas and enclaves where there is too little traffic to make*
> *new infrastructure viable, delay in providing infrastructure means*
> *that these regions cannot be properly linked in. (p8)*

These imbalances, proponents contend, occur when transport markets do not function properly. Poor organization and management of transport systems, the European Commission (2001) argues, has resulted in distorted transport prices. Lax enforcement of safety and environmental standards on the roads have meant that the price of road transport does not accurately reflect its costs (European Commission, 2001). This, the Community of European Railway and Infrastructure Companies (CER) contends, is exacerbated by unfair taxation regimes that favour road transport over rail (CER, 2007). Such prices, proponents contend, offer poor incentives for choosing cleaner, safer or less congestion-prone means of transport (European Commission, 2001). Additionally, it would seem as if individuals cannot be trusted to make rational decisions about car use: despite the well-publicized fact that it affords least safety to users, the car remains the most popular choice of transport (European Commission, 2001). The upshot, proponents of the balanced mobility story contend, is that in Europe distorted prices and user myopia have favoured road transport to the detriment of other modes, particularly the railways (European Commission, 1993, 2001).

These imbalances are the reason why providing transport infrastructure alone will not work (European Commission, 2001; Department for Transport, 2004). Because transport prices are distorted, infrastructure investment alone will worsen modal and regional imbalances. This, in turn, incurs prohibitive environmental and social costs (European Commission, 2001; Department for Transport, 2004). Consequently, the European Commission (1995) concludes that ...

> *... [r]educing pollution and congestion by means of increasing road*
> *capacity is – in many cases – not the best option. The cost of con-*
> *struction of road (and parking) capacities in densely populated areas*
> *continues to increase. Studies indicate that improving and extending*
> *infrastructure results in more journeys overall as road users make use*
> *of the new or improved facilities. The environmental impacts both of*
> *these extra journeys and of the construction of the road infrastruc-*
> *ture may outweigh any benefits in improved traffic flows. (p4)*

In other words, building more roads simply defeats the object. The Department for Transport (2004) summarizes the transport policy challenge as follows: 'We cannot build our way out of the problems we face on our road networks. And doing nothing is not an option' (p14).

The heroes: uncoupling transport and economics

How are transport policy-makers to avoid the Scylla of congestion without getting caught by the Charybdis of environmental and social degradation?

If policy-makers are to foster economic growth without compromising the natural or social worlds, proponents of the balanced mobility story contend, then transport policy-making must disconnect transport from economic growth. By '... gradually breaking the link between economic growth and transport growth ...' (European Commission, 2001, p11) transport policy-makers can construct '... a transport network that can meet the challenges of a growing economy and the increasing demand for travel, but can also achieve our environmental objectives' (Department for Transport, 2004, p12).

Restoring equilibrium to transport networks is definitely a hands-on task. The transport policy issue, the European Commission (2001) muses, is a '... complex equation that has to be solved in order to curb transport demand' (p10). Solving these equations effectively, proponents contend, calls for an even-handed policy strategy (Department for Transport, 2004). Balancing divergent concerns in increasingly fluid policy environments requires 'a broader, more flexible transport policy toolbox' than has been available in the past (European Commission, 2006, p10). For solutions to contemporary transport problems to stick, so the European Commission (2001) contends, policy-makers need to implement strategies consisting of '... a series of measures ranging from pricing to revitalizing alternative modes of transport to road and targeted investment in the trans-European network' (p11).

Given the projected growth of transport demand, one central policy tool will have to be investment into transport infrastructure (European Commission, 1993, 2001, 2006; Department for Transport, 2004; CDU, 2007). As part of their 'balanced approach' to transport strategy, Tony Blair promises that where '... it makes economic sense, and is realistic environmentally, we will provide additional transport capacity' (Department for Transport, 2004, p5). Infrastructure investment in railways is also a way of redressing the imbalance between the road and other modes of transport (European Commission, 2001; CER, 2007).

But, as we have seen, building roads, extending railways and upgrading ports is only part of the solution. Balancing economic growth with sustainable environmental and social development will require changing the nature of transport demand itself. Since distorted transport prices have skewed demand towards cars and lorries, policy-makers need to implement measures to include external costs in current transport prices. For this reason, the CER (2007) is ...

> ... *convinced that the internalization of external costs is the primary instrument to be implemented in order to create a level playing field between transport modes; that is definitely promoting the competitiveness of transport modes and consequently sustainable transport strategy. (p2)*

These so-called market-based-instruments (MBIs) include forms of road or energy taxation as well as road charging and toll schemes (European Commission, 2001; Department for Transport, 2004; Banks et al, 2007).

Regulation is another important instrument in the transport policy toolbox. For example, the fragmentation of air traffic control in Europe ...

> ... *adds to flight delays, wastes fuel and puts European airlines at a competitive disadvantage. It is therefore imperative to implement, by 2004, a series of specific proposals establishing Community legislation on air traffic and introducing effective cooperation both with the military authorities and with Eurocontrol. (European Commission, 2001, p14)*

Similarly, the SPD in Germany plans to avoid unnecessary mobility by more intelligent land-use and logistics regulation.

Last, as unpopular as they may be in an era of 'modern governance', classic command-and-control measures, when used wisely, are effective transport policy tools. In cases where policy-makers cannot rely on the common sense of transport users, policy may need legislative and executive means to pursue policy objectives. ASECAP (2002) contends that despite ...

> ... *the intrinsic qualities of motorway infrastructures, in most cases accidents are caused by the inappropriate behavior of drivers. It seems that controlling the speed of vehicles, whether to improve safety, reduce congestion or decrease polluting gas emissions, is now a necessity from which infrastructure managers cannot escape. (p10)*

But, so proponents of balanced mobility argue, process is as important as substance. In decoupling transport demand from economic growth, transport policy-makers need to take great care: the European Commission (2001) contends that to ...

> ... *take drastic action to shift the balance between modes – even if it were possible – could very well destabilize the whole transport system and have negative repercussions on the economies of candidate countries. Integrating the transport systems of these countries will be a huge challenge to which the measures proposed have to provide an answer.*

Moreover, balancing transport systems implies that transport policy-makers integrate different policy tools into a coherent and effective strategy. Effectively deploying infrastructure investment, market-based instruments or regulation means understanding the impacts on the economy, the environment and society (European Commission, 2001; Department for Transport, 2004). On this view, environmental and social parameters must become part of the economic calculus. By the same token, environmental and social policy measures need to take into account profitability and economic growth. This, advocates of balanced mobility contend, requires in-depth knowledge of the way different transport modes operate and interact with other social systems.

Table 3.1 depicts the scope of policy conflict in the European Transport Debate.

The structure of policy conflict in transport policy: agreement and disagreement

The scope of conflict in the European transport debate shows that conflict is endemic. However, this is not to say that there is no agreement. The conceptual framework developed in Chapter 2 suggests that the three types of inherently antagonistic policy frames also share some concerns across advocacy coalition borders. In terms of the grid/group diagram, each social solidarity shares the same dimensional space with another social solidarity: hierarchical and egalitarian advocacy coalitions have in common a cohesive organizational structure while egalitarianism and individualism are both characterized by a lack of internal organizational distinction and stratification. The conceptual framework would lead us to expect some overlap on basic principles and general policy measures. This agreement, however, is inherently fragile because it derives from fundamentally divergent premises.

How does this topography of agreement and disagreement map onto the European transport policy debate?

Table 3.1 *The scope of policy conflict in the European transport debate*

	Efficient mobility	*Sustainable mobility*	*Balanced mobility*
Setting	There is a close relationship between mobility, opportunity and prosperity; policy-makers need only make sure that the virtuous mobility–opportunity–prosperity spins freely; this means (a) providing more and more effective transport infrastructure (mostly roads) and (b) deregulating transport markets to spur innovation	Modern transport systems distribute costs and benefit inequitably; In order to appreciate these injustices, we need to adopt a holistic view of transport policy	Transport is vital for economic growth and our way of life; modern transport systems have adverse impacts on the environment and society that threaten to undermine our high quality of life; transport policy is about balancing benefits and costs of modern transport; hands-on management task
Villains	Congestion is slowing our societies by strangling dynamic economic growth; congestion is the result of misguided policies that aim to interfere with transport markets by favouring inefficient modes (i.e. rail); much of this unfair and wrong-headed policy is based on dubious theories and weak evidence; transport policy is driven by irrational politics	Modern transport systems are destroying natural and social environments; road transport contributes to environment degradation; transport infrastructure is destroying peripheral communities in favour of economically powerful regions; modern transport systems systematically disadvantage weak and vulnerable; the mobility–opportunity–prosperity cycle is a widespread myth	Transport systems are imbalanced; on the one hand, there is a lack of infrastructure and on the other hand, road dominance means that simply building more roads leads to more congestion and pollution; transport markets are imperfect and create large externalities
Heroes	Provide effective and efficient transport infrastructure; allow innovative entrepreneurs to find efficient solutions to environmental and social impacts of transport; technological innovation will solve environmental problems (which have been largely exaggerated anyway)	Fundamental reorganization of economic order along ecological and sustainable lines is needed; Introduce new forms of land-use management; radically reduce demand for transport in general and road transport in particular; Internalize external costs; develop sustainable alternatives to current transport choices; get rid of inequities in the transport system	Uncouple transport demand from economic growth; breaking the link needs to be managed carefully; use a wide range of policy instruments (road charges, management measures, infrastructure investment); it needs wide meta-governance of transport policy

Areas of agreement: broad principles, common policy measures and mutual rejection

Comparing the contending policy stories reveals areas of pairwise agreement and disagreement within the contested terrain. While all advocacy coalitions agree that transport policy is in some way crucial for European societies, more specific policy principles and policy measures are shared by two of the three contending advocacy coalitions. Furthermore, each advocacy coalition shares a distaste for the specific policy measures with another advocacy coalition.

Proponents of the efficient mobility and balanced mobility stories agree that transport systems and transport policy significantly impinges on economic growth. As we have seen, members of these advocacy coalitions both assume that economic growth is a central, possibly the central, policy goal for any government. Therefore, both advocacy coalitions agree on policy measures that increase the efficiency of existing transport networks: policies to cut average trip times and costs per trip find the support of members from both the hierarchical transport planning approach and the individualist market-oriented policy perspective. This can, in principle, include investment in existing and new infrastructure, road pricing schemes or telemetric traffic management systems. Conversely, both the hierarchical and individualist advocacy coalition take a dim view of policy measures aimed at reducing mobility.

Advocates of both the sustainable mobility story and the 'balanced mobility story' share a concern for the integrity of the natural and social environment. As we have seen, advocates of balanced mobility see transport systems as the primary means for securing, through regional economic growth, social and regional balanced mobility. This growth, however, needs to be tempered by and set in the framework of 'sustainable development': both agree that modern transport systems will have to uncouple economic growth from the growth in transport demand. Similarly, sustaining and preserving this micro-mobility, in turn, goes hand-in-hand with preserving and protecting the natural environment. For this reason, members from both advocacy coalitions support transport policies that protect social and natural environments. By the same token, both hierarchical and egalitarian advocacy coalitions reject policies that threaten human and natural habitats. Measures to deregulate and liberalize freight and transport markets, then, attract opprobrium from these advocacy coalitions.

Last, the efficient mobility and sustainable mobility stories share a preference for decentralized transport management. Members of both advocacy coalitions believe that decision-making in transport policy and transport choices ought to be taken at the most suitable level of governance.

It is exclusively at this level that individual and human rights can be best protected and fulfilled. For this reason, members from both advocacy coalitions favour policy measures that decentralize and devolve decision-making away from central state bureaucracies. It follows that the egalitarians and the individualists are united in their opposition to large, centralized and inflexible transport systems. Projects such as the Trans-European Networks or high-tech, prestige projects such as the German Transalpin attract the criticism from both advocacy coalitions.

Intractable disagreement

The agreement on basic principles and policy measures in the transport policy domain evaporates once debate moves away from the general level. While the advocacy coalitions are clear on what they reject, their rejection is driven by divergent motivations. At a more concrete level, advocacy coalitions in the transport debate fail to agree on the definitions of the issue, the causality of the problem or even the approach for understanding the phenomenon.

Efficient vs balanced mobility

Although proponents of the efficient mobility and balanced mobility stories argue that transport systems affect economic growth, they disagree on *how* the two are related. As we have seen, members of the individualist advocacy coalition see a direct and positive relationship between infrastructure investment, mobility and economic growth: more of one inevitably means more of the others. In contrast, members of the hierarchical advocacy coalition have become sceptical of the basic assumption of classic location theory. For example, Peter Hall points out that the relationship between accessibility and development is 'one seamless web'. However, he contends that the '... trouble is that this combination is so subtle, no one anywhere seems to have completely understood how to make it work at a fine-tuned level' (Hall, 1995, p75). Himanen et al (1993) argue that, like the sorcerer's apprentice, transport experts helplessly face the transport demons they have summoned. The uncertainties in transport planning theory and methodology are so significant, Banister and Lichfield are forced to admit that the ...

> ... *precise relationship between transport investment and urban development is not well known, even theoretically. There seems to be no single methodology available to test the relationships, the counterfactual situation is difficult to determine and the question of causality not addressed. Decisions have been made based more of*

faith than understanding. Even where clear methodological approaches have been tried, problems arise concerning available data and the inherent complexity of many relationships. The links between land use, transport and development are much more profound than just an examination of the physical, social and economic relationships might produce. (Banister and Lichfield, 1995, p15)

Sustainable vs balanced mobility

The agreement between the proponents of the sustainable mobility and balanced mobility stories also dissipates on closer inspection. Although both advocacy coalitions are in favour of conserving natural and social environments, they conceive of these ideas in very different terms. Sustainable development for hierarchical advocacy coalitions in the transport policy debate is about factoring in sound management and conservation principles to transport existing planning paradigms. The egalitarian policy story contends that the integrity of natural and social environments is a fragile and priceless thing that, once destroyed, is inordinately difficult to reconstruct (Knoflacher, 1997). Hence, transport policy must prevent the destruction of natural and social habitats at all costs.

The main problem, advocates of sustainability argue, is that the transport planners have no way of even appreciating or understanding the true value of social and natural environments. The egalitarian policy story discredits the entire conceptual edifice as a poorly disguised legitimation for existing inequities. Starting with the assumptions and ending with the conclusions, so the argument goes, the entire analytical architecture is rigged to produce the 'correct' results, which means results that objectively prove the inherent superiority of large-scale, complex and capital-intensive infrastructure investment. Indeed, as champions of sustainability point out, while some transport policy-makers have taken on board environmental arguments, they still are a long way from adopting a truly holistic view of transport policy. This is why T&E (2003) laments that the European transport policy debate is too narrow: 'Economics is thoroughly discussed and environmental issues are well-publicized. But transport's social problems are often unclear and the social/psychological factors supporting transport patterns are often forgotten' (p1).

Efficient vs sustainable mobility

Last, despite the shared distaste for unwieldy centralized transport management, both the sustainable mobility and efficient mobility story have little more in common. Both advocacy coalitions favour decentralization for diametrically opposed reasons. Individualists' advocacy coalitions want to see allocation and supply decisions in the transport returned to the level of the individual economic agent (i.e. individuals and firms) in order to maximize accessibility and traffic flow. Conversely, the egalitarian policy story maintains that decision-making needs to be returned to the local and community level in order to restrict accessibility and flow of traffic through the system.

For proponents of the sustainable mobility story, the individualist idea of decentralizing transport decision-making is rather thin. The basic problem, contend proponents of sustainable mobility, is that advocates of mobility conceive of human rights as nothing more than narrow consumer rights. Indeed, the root of all problems, they argue, is the fact that the inequitable, unsustainable and unjust socio-economic order can only conceive of people as one-dimensional consumers. Worse still, this system judges people solely in terms of the amount they consume. Ever faster and bigger cars, larger, more impressive houses in the leafy suburbs or luxurious holidays in exotic places on the other side of the world have, absurdly, become measures of our values as human beings. While trying to keep up with the Joneses – more often than not in vain – we have lost touch with our real human needs: living autonomously without fear of oppression in healthy natural and social environments. That is why, T&E (2003) argues, that movement ...

> *... may be a right, but is not a 'trump card' with which to justify thoroughly unsustainable behaviour. It's a well-established principle that rights are limited: for example, the right to free speech does have limits. So, although people have a general right to physical mobility, the social (and other) consequences of how they exercise this right are very important, and it has limits (rights of others to more basic rights, like health). (p3)*

In sum, a 'contested terrain' featuring hierarchical, individualist and egalitarian types of advocacy coalitions from the cultural theory typology of social solidarities will give rise to endemic policy conflict. This is, in part, due to the structure of policy conflict, that is the way contending policy frames carve out areas of agreement and disagreement. In the European transport policy debate, agreement on basic principles and general policy measures as well as mutual rejection spans two of the three contending advocacy coalitions. However, since all three policy frames discussed are grounded in fundamentally divergent forms of organization, they converge towards the areas of agreement from opposite directions. Divergent issue definitions and underlying theories of causality mean that agreement is limited and restricted to a highly general level of analysis: debate of a more concrete type unearths more fundamental and intractable differences.

Table 3.2 summarizes the areas of agreement and disagreement between the contending advocacy coalitions.

Table 3.2 *Structure of conflict about transport*

	Efficient mobility – Balanced mobility	*Sustainable mobility – Balanced mobility*	*Efficient mobility – Sustainable mobility*
Agreement	Transport policy significantly impinges on economic growth	Concern for integrity of natural and social world	Decentralized transport systems
Mutual rejection	Measures aimed at reducing mobility	Measures to deregulate freight and transport markets	Large, centralized and inflexible transport projects (e.g. German Transalpin)
Disagreement	Mechanism that relates transport to economic growth	Transport policy tools and approaches	Shape and form of decentralized transport

The scope and structure of policy conflict, then, create a triangular policy space that delimits the boundaries of the contested terrain (see Figure 3.1).

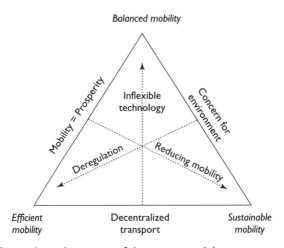

Figure 3.1 *Triangular policy space of the transport debate*

The potential impacts of policy conflict: weaknesses and conceptual blindness

The fact that all policy knowledge is contextual and that transport policy debate is inherently conflictual does not mean that constructive policy debate is impossible. Policy frames are not irrational nor are policy stories inaccurate: each policy frame merely foregrounds certain issues and backgrounds others.[1] That is why each policy frame captured by the cultural theory typology has its strengths and weaknesses.

This is not to say, however, that the road to effective transport policy-making lies in uncritically accepting any form of analysis. Just because we cannot conclusively prove any one policy frame to be 'wrong' does not mean that all diagnoses and policy recommendations must be 'right'. Since we must distinguish frames and policy stories not by what they include into their analytic frameworks but by what they exclude or filter out, policy frames create cognitive and perceptual 'shutters' or 'conceptual blindness'. Policy processes that rely on one frame alone (or alternatively do not encourage serious criticism from proponents of contending advocacy coalitions) introduce a systematic bias into their analysis of transport problems: this, as Collingridge[2] has demonstrated, can lead to myopia, poor analysis and, ultimately, poor policy decisions.

Efficiency and market failure

The central weakness of the efficient mobility story is its unswerving trust in the market. The mobility–opportunity–prosperity cycle assumes that, at any time and in any place, markets allocate resources more efficiently than any other distributive system. On this view, markets fail for one and only one reason: someone – typically government – has been meddling with the unerring self-equilibrating and self-regulating mechanism of the invisible hand. Yet, there are many reasons, apart from government interference, that may cause markets to fail. Like all other forms of social cooperation, markets depend on a favourable institutional and environmental framework. Markets may fail because prevailing socio-cultural values are unfavourable to market-type transactions (for example, the belief in equivalent exchange values). Alternatively, markets may collapse because transaction costs unrelated to production or marketing costs are prohibitively high.[3] Another reason for market failure may be an insufficient legal and administrative framework that can effectively implement sanctions. There are, in short, a plethora of cultural, social and political reasons for market failure: the catch-all phrase of 'government interference' is either too all-inclusive to provide an adequate grasp of all possible sources of market failure or it is too limited if it refers to government intervention alone. Consequently, for champions of mobility, the solution to market failure is to call for more market.

Since the efficient mobility story has little appreciation of institutional path-dependencies, technological lock-in and social structuration, the solutions it proposes are vulnerable to creating and exacerbating the types of disparities in transport systems flagged by proponents of the sustainable mobility story. Transport markets never operate in a socio-economic or technological vacuum. Despite years of welfare state and social mobility, European societies are still predominantly class societies characterized by pervasive inequalities. What is more, transport markets trade within entrenched technological trajectories. Change in either social structures or technological trajectories is rather costly. Competitive transport markets, then, favour those with the wherewithal to compete. In societies characterized by inequities and technological lock-in, comparative advantages may be unjustly and inequitably distributed. In other words, rather than rewarding hard work, innovation and entrepreneurial spirit, transport markets in inequitable societies may simply entrench the dominant position of the monied classes and large-scale industries, most notably automobile and aeronautics.

Sustainability and intolerance

The sustainable mobility story, in turn, is overly confident of the inherent moral superiority of their argument. First, the advocates of the egalitarian ecological approach claim to understand peoples' real human needs better than individuals do themselves. Rather than the false needs that the corrupt system imposes on humans, such as a desire for material goods, striving for status, or conforming to social pressures, the ecological view points to real human needs: life in harmony with the natural and the social world. Unlike the other approaches, the egalitarian ecological perspective aims not only to change societal structures but also to alter the way individuals think. Within the ecological view there is a definite moral 'right path'; other modes of thinking are polluted, corrupt, and immoral.

Second, many advocates of the sustainable mobility story are obsessed with pollution – both environmental and moral – from a corrupt outside world. In the extreme, this takes a rather xenophobic form. For example, Knoflacher depicts transit traffic through the Austrian alpine regions as little more than '… an alien body in the organism of the state. A healthy body will, if it is to survive, reject all alien bodies (*Fremdkörper*)' (Knoflacher, 1997, p121). The upshot of his argument is that regional and local freight and passenger transport is justified because it is 'life-sustaining' to a particular community. Conversely, transit transport, which benefits those outside a community, is an evil to be eradicated.

This is not to say that all proponents of the sustainable mobility story are inherently xenophobic (cf. Hillman, 1999 or Adams, 2005). Yet an integral part of the argumentative logic is an extreme distrust and suspicion of proponents of other policy arguments and frames. This is reflected in the elaborate and indiscriminate conspiracy theories this view constructs to explain transport policy problems. The unholy alliance of big business, corrupt government and dishonest researchers conspires to maintain adequate profit levels at the cost of human well-being (Adams, 1995, 2005; Knoflacher, 1997). What is more, not only has big business captured government agencies and university departments, it also has succeeded in poisoning the minds of most people (T&E, 2003).

All this tends to generate a degree of intolerance towards other modes of thought. In particular, advocates of the sustainable mobility story have little respect for what morally corrosive political and economic systems call civil liberties. Knoflacher states that the freedom of mobility as well as the freedom of settlement contravene real human needs such as a right to a healthy life, a right to personal development and a right to a home '*Recht auf Heimat*' (Knoflacher, 1997).

Although there can be no doubt that transport policy-making in the past and at the moment is significantly, and perhaps unduly, influenced by commercial interests,[4] sweeping generalizations of the Knoflacherian type are inaccurate. Closer inspection of transport policy-making reveals that, like all policy spheres in advanced capitalist states, the state is fragmented, fissiparous and highly specialized. In short, transport policy-making also takes place in the 'differentiated polity'. The sheer number and variety of policy actors in these networks undermines the credibility of any general conspiracy theory. Often, government agencies in different policy networks will pursue contradictory policy goals. What is more, presently there is little evidence of iron triangles in transport policy-making. For example, the European Commission, often regarded as the lackey of big business interests by the proponents of the sustainable mobility story, is forcefully pushing road pricing policy options which run counter to the economic interests of large-scale commercial enterprises (European Commission, 1995a).

Balance and congestion

Last, the balanced mobility story places an inordinate amount of trust in the managerial capacities of transport planners. At a more basic level, champions of balanced mobility believe that the transport policy issue is fundamentally manageable. Proponents of the balanced mobility story understand the messiness of the transport policy issue as a complicated equation with a large number of unknowns. They will freely admit that solving this equation is a difficult task: this is precisely why not everyone is cut out for the job. Only experts with the requisite technical skills and balanced judgement ought to be allowed. What proponents of the sustainable mobility story see as an issue of equity and the advocates of efficient mobility perceive as a question of individual liberty is really a far more mundane issue of optimization, regulation and control. Mundane, that is, if you have the requisite skills of measurement and assessment. Solving complicated transport policy conundrums, then, is about finding or developing the right technical means of control. This invariably leads to the search for an ever bigger and technically sophisticated transport model with which to measure socio-economic and ecological impacts of transport infrastructure investments.

But what are transport planners measuring? To take the SASI transport model (Box 3.5) as one example of many, the model purports to provide indicators for regional balanced mobility and thus allows policy-makers to assess the socio-economic impacts of transport investments. However, the model reduces regional development and balanced mobility to GDP per capita and regional unemployment respectively: as we have seen, Bökemann et al (1997) seem to believe that this covers economic as well as social impacts. At best, however, these indicators measure an economic impact of transport infrastructure investment; there is no social dimension to these indicators whatsoever. The indicators leave income distribution, the access to public and private services and the socio-cultural fabric, to say nothing of social exclusion, ethnic and gender issues completely untouched. Surely, indicators of regional balanced mobility should have something to say about horizontal social, cultural and political ties between two locations. Yet all the SASI indicators tell us is whether transport infrastructure investment has succeeded in harmonizing regional unemployment rates which really indicate the degree of social exclusion within a community.

Box 3.5 *The SASI model*

Planning and allocation decisions in transport policy-making heavily rely on quantitative models based on discounting techniques such as cost–benefit analysis (CBA) and multi-criteria analysis (MCA). The goal of transport models is to '... measure the impacts of transportation policies in general and specific transportation system investments in particular' (Bökemann et al, 1997, p9). Exact measurement and quantification of the socio-economic impacts associated with increased accessibility is necessary if planners are to assess the degree to which accessibility fosters economic growth. Furthermore, given the volume of expenditure in transport projects such as '... the Trans-European Networks (TETN) programme, the need for consistent prediction and rational and transparent measurement of the likely socio-economic impacts of major transport system improvements becomes obvious' (Bökemann et al, 1997, pp8–9). Three things are important here: first, the belief that transport systems are controllable; second, that complex transport phenomena are amenable to quantitative modelling and prediction; third, that rational models imbue the transport planner with control over highly complex social, economic and political processes.

How do planners know that they have fulfilled their objectives? In other words, how do planners define accessibility in the concrete terms of quantitative indicators? Again, the SASI project provides a good example of how planners translate the basic assumptions into functional model parameters. The SASI (socio-economic and spatial impacts of transport infrastructure investments and transport systems improvements) project aims to measure both the social and economic impacts of transport infrastructure investment as well as its contribution to European regional balanced mobility. Since accessibility means economic growth, the only logical indicator for the degree of accessibility is regional gross domestic product per capita (Bökemann et al, 1997, p12). Regional cohesion in turn, they argue, is a more difficult concept to quantify: that is why '... the cohesion measures used in this study reflect the even or unevenness of the spatial distribution of socio-economic and accessibility indicators among the regions' (Bökemann et al, 1997, p34). There are, then, three sets of indicators to cover both the social and the economic aspects of accessibility. The economic measure of accessibility is, as mentioned, gross domestic product per capita. Regional unemployment, in turn, measures the social availability of jobs. Last, the SASI project uses both income and regional unemployment as cohesion indicators on the European scale.

These indicators reflect the basic assumptions of the classical planning view. In the SASI model, accessibility is a pointer to regional gross domestic product: the more accessibility, the higher the gross domestic product. The qualitative dimensions of accessibility, such as the social exclusionary effects of poor public transport, low car ownership, etc. have been faded out of analytic perspective. In sum, the set of indicators is tied to a structural conception of accessibility.

What is more, the classical transport planning approaches systematically deny the political nature of transport models. Cloaked in the veil of science and objectivity, transport planners maintain the illusion that they stand aloof from the political fray of transport decision-making. Rigid methodology, strict quantification and hard data ensure, so the story goes, that research is objective. Transport planning methods do not explicitly provide policy options: they merely provide the 'facts' and let these speak for themselves. Yet, as we saw in Chapter 2, facts tend to be a taciturn lot. By insisting that policy be informed by 'objective' analysis and then proceeding to define what 'objective analysis' entails, the transport planning approaches a priori disqualify any alternative form of analysis. Defining the transport policy issue as a technical problem amenable to technical solutions precludes any other approaches.

The strength of the balanced mobility story is also its greatest weakness. Transport planners are very adept at formalizing, quantifying and modelling certain transport processes. However, the strictures formal models impose on researchers causes transport planners to spectacularly miss the point. By excluding all factors not amenable to exact and objective measurement, transport planners invariably fail to grasp the inherent complexity of transport policy processes. Small wonder that the world seems manageable when transport modellers conveniently ignore the things that make transport policy messy. As a result, most transport models fail to produce conclusive results even within their highly restricted analytical scope.

We can see that if policy-making were in the metaphorical hands of one advocacy coalition alone, we would be left with one-sided policy solutions. The proponents of the efficient mobility story would be busy constructing transport infrastructure without regard to social or environmental effects. The advocates of sustainability would implement policy solutions that favour small-scale local activity regardless of the wider economic and social implications. The proponents of balanced mobility would build elaborate regulatory regimes to uncouple transport demand from economic growth that would ignore the very things that wed the two together.

Table 3.3 summarizes the potential impacts of each policy story.

Table 3.3 Potential impacts of policy conflict

	Efficient mobility	**Sustainable mobility**	**Balanced mobility**
Trusts	innovative capacity of transport entrepreneurs	inherent superiority of their moral position	managerial capacities of transport experts and their tools
Downplays	inherent market distortions and adverse impacts on environment	self-interest of actors	political nature of transport models
Vulnerable to	increasing transport disparities	intolerance and zealotry	models that may miss the point and ignore the things that makes the transport messy

Conclusion

Different groups of policy actors resolve the transport dilemma in very different ways. These divergent approaches to transport policy in Europe, however, are not random or irrational. They merely reflect the differing policy frames through which policy actors interpret the uncertainties and complexities involved in contemporary transport systems. In order to explore these policy frames and their significance for European transport policy-making, this chapter has applied the framework developed in Chapter 2 to the transport policy debate.

Using cultural theory's typology of policy frames, the chapter sounded out the scope of policy conflict in the 'contested terrain' of transport policy in Europe. The analysis revealed three competing policy stories in the transport debate. The first, a market-oriented discourse, understands mobility to be the motor that drives dynamic and innovative societies. Anything that slows this motor, such as congestion resulting from government meddling with transport markets, will slow societal progress as a whole. Solving the transport policy dilemma means removing obstacles to mobility by providing transport infrastructure to meet burgeoning demand. The sustainable mobility story argues that modern transport systems, integral parts of the unjust and exploitative socio-economic and political order, have destroyed our environments so that they no longer satisfy real human needs. On this view, present transport dilemmas are merely a symptom of a fundamentally flawed, inequitable and misanthropic socio-economic system; solving the dilemma means replacing the system with something more sustainable. Last, the balanced mobility story recognizes both sides of the argument: while transport systems are vital for our prosperous way of life, they also generate environmental and social pressures that jeopardize that very way of life. The solution to this dilemma, so the argument goes, does not lie in either extreme. Rather, transport policy-makers need to balance these potentially contradictory forces by judicious and skilful transport management and planning.

These three policy stories delineate the boundaries of the contested terrain in the European transport policy debate: they both enable and constrain thinking and argument about transport policy. Since policy frames are inherently normative (they legitimate fundamentally divergent forms of social organization), and since transport is a messy policy, conflict in transport policy debates is inevitable and endemic. The contending policy stories set up the *structure* of this policy conflict by carving out areas of agreement and disagreement between the advocacy coalitions. In a contested terrain like the European transport policy domain, agreement is likely to be fragile because, on the one hand, it fails to hold beyond basic principles and general policy measures and, on the other hand, it is fundamentally limited since common ground is accessible to only two advocacy coalitions. At more concrete levels, agreement collapses under the weight of the incommensurable assumptions and modes of perception that divergent advocacy coalitions bring to the debate. Science, research and 'technical knowledge' are unlikely to help resolve this conflict: as we have seen, advocacy coalitions also fundamentally disagree about what is to count as a fact and what is to count as legitimate knowledge. Given this constellation of discourses and advocacy coalitions in the contested terrain, a policy subsystem featuring two or more advocacy coalitions would be likely to give rise to an 'intractable policy controversy'.

Thus, the discourse analysis would seem to suggest that transport policy-making would be much improved if it were controlled by one advocacy coalition. But which one? Since each policy frame is informed by a specific cultural bias, policy stories invariably tell an incomplete and selective tale. Each type of advocacy coalition and corresponding policy frame is characterized by specific weaknesses and strengths. Leaving one coalition and its story alone at the transport policy-making helm, then, may lead to policy failures. Although a wide scope of policy conflict increases the likelihood of a 'dialogue of the deaf', it also generates the widest reservoir of ideas, concepts and strategies for tackling the European transport policy issue.

What, then, are the implications of this analysis for the European transport policy debate?

- Policy conflict is endemic. Policy actors rely on incommensurable policy frames to make sense of the transport policy dilemma. The contending policy stories about transport start from divergent assumptions, identify different causes and propose conflicting policy solutions. This policy conflict is unlikely to resolve in the face of facts and objective knowledge because the transport debate thematizes the validity and legitimacy of different forms of knowledge about transport policy-making.
- The contending policy stories set up tenuous spaces of agreement on general principles and policy measures limited to pair-wise alliances of advocacy coalitions. This agreement is fragile for two reasons. On the one hand, the imbalanced structure of agreement implies that an advocacy coalition is always excluded from consensus. On the other hand, pairwise agreement is limited to general principles and policy measures; it breaks down into an intractable controversy at more concrete levels.
- Since the likelihood of unanticipated consequences and policy failure increase with decreasing accessibility of the policy subsystem, policymakers should enable a lively policy debate with as many advocacy coalitions as possible. However, given that accessibility inevitably leads

to value-driven policy conflict, policy-makers need to ensure that the quality of communication and the responsiveness of the debate do not deteriorate in the course of policy debate.

Notes

1 In fact, policy frames, if they are to survive scrutiny, must be credible. Credibility, in turn, has a lot to do with the predictive and explanatory powers of a particular approach, see Wynne and Jasanof, 1998.
2 Collingridge reviews cases such as the Space Shuttle disaster, high-rise building schemes in the UK and nuclear power. Unfortunately, he does not address transport, although the analysis and conclusions of his case studies equally apply to transport policy-making (Collingridge, 1992).
3 See the literature on common pool resource (CPRs). Here, transaction and monitoring costs are sufficiently high – as are the incentives to renege on contracts – that efficient market transactions are impossible (Ostrom, 1990).
4 For an enlightening, if frustrating, insight into the relationship between stakeholder consultation and transport policy outcomes in the south-west of the UK, see Kinnersly (2001).

References

ACEA (2002) 'The ACEA Response – European Transport Policy for 2010: Time to Decide', www.acea.be/index.php/files/the_european_commissions_white_paper_on_european_transport_policy_for_2010_, accessed 2 October 2008

ACEA (2007) 'Commission's Public Consultation on the Preparation of a Green Paper on Urban Transport – Cities Need Sustainable Efficient Mobility: a View from the Automotive Industry', www.acea.be/index.php/files/urban_transport_acea_input_to_the_european_commissions_green_paper, accessed 7 October 2008

Adams, J. (1995) *Risk: the Policy Implications of Risk Compensation and Plural Rationalities*, UCL Press, London

Adams, J. (2005) 'New Modes of Governance: Developing an Integrated Policy Approach to Science', *Technology, Risk and the Environment*, Ashgate, Aldershot

ASECAP (2002) 'Opinion of the European Concession Motorways Sector regarding the White Paper of the European Commission on the Transport Policy until 2010: 'A time for choices', www.asecapcom/english/pubinf-posofficielles-en.html, accessed 5 January 2008

Balaker, T. (2006) 'Why Efficient Mobility Matters', *Reason Foundation Policy Brief*, no 46, Los Angeles, CA

Banister, D. (2008) 'The Sustainable Efficient Mobility Paradigm', *Transport Policy*, vol 15, November, pp75–80

Banister D. and Button, K. (1993) 'Environmental Policy and Transport: An Overview' D. Banister (ed) *Transport, the Environment and Sustainable Development*, E&FN Spon, London

Banister D. and Lichfield, N. (1995) 'The Key Issues in Transport and Urban Development' in D. Banister (ed) *The Key Issues in Transport and Urban Development*, E&FN Spon, London, pp1–15

Banks, N., Bayliss, D. and Glaister, S. (2007) 'Motoring Towards 2050: Roads and Reality', *RAC Foundation Research Report*, London

Bökemann, D., Hackl, R. and Kamar, H. (1997) 'Socio-Economic and Spatial Impacts of Transport Infrastructure Investments and Transport System Improvements', *Socio-Economic Indicators Model Report – EUNET Project*, Vienna

Bündnis 90/Die Grünen (2002) 'The Future is Green: Party Program and Principles', www.gruene.de/cms/files/dokbin/145/145643.party_program_and_principles.pdf, accessed 30 September 2008

CDU (2007) (Christlich Demokratische Union Deutschlands) 'Deutschland braucht eine moderne, zukunftsweisende Verkehrsinfrastruktur', www.cdu.de/politikaz/verkehr.php, accessed 9 January 2008

CER (2007) Green Paper on Market-Based Instruments for Environment and Related Policy Purposes, *Position Paper*, SEC(2007) 388

CLECAT (2005) *CLECAT Response to Review of 2001 White Paper on European Transport Policy, Position Paper*, Brussels

Collingridge, D. (1992) *The Management of Scale: Big Organisations, Big Decisions, Big Mistakes*, Routledge, London

Department for Transport (2004) 'The Future of Transport: a Network for 2030', *CM 6234*, London

European Commission (1993) 'White Paper on Growth, Competitiveness, and Employment: The Challenges and Ways Forward into the 21st Century', *COM(93) 700 Final*, Brussels

European Commission (1995) 'Towards Fair and Efficient Pricing in Transport: Policy Options for Internalising the External Costs of Transport in the European Union', *COM(95) 0691 Final*, Brussels

European Commission (2001) 'European Transport Policy: Time to Decide', *COM(2001) 370 Final*, Brussels

European Commission (2003) 'European Energy and Transport Trends to 2030', *Research Report*, Brussels

European Commission (2006) 'Keep Europe Moving – Sustainable Efficient Mobility for our Continent: Mid-term Review of the European Commission's 2001 Transport White Paper', *COM (2006) 314 Final*, Brussels

European Council of Ministers of Transport (1998) 'Statistical Trends in Transport 1965-1994, ECMT, Paris

European Transport Workers' Federation (2004) 'Statement on Social Consequences of European Transport Infrastructure Policy', http://www.itfglobal.org/files/extranet/-75/360/ETF%20Statement%2014_06_04.pdf, accessed 4 March 2008

EEA (2006) (European Environmental Agency) 'Transport and Environment: Facing a Dilemma', EEA Report, no 3/2006, Copenhagen

ETSC (2002) (European Transport Safety Council), 'ETSC Response to the Commission's White Paper "European Transport Policy for 2010: Time to Decide"', www.etsc.be/documents.php?did=3, accessed February 2008

ETSC (2005) (European Transport Safety Council), 'ETSC response to the Commission's Consultation document on the White Paper', www.etsc.be/documents.php?did=3, accessed 6 February 2008

EUROSTAT (2007) *Panarama of Transport, 1990–2005, Fifth Edition*, Office for Official Publications of the European Communities, Brussels

Hall, P. (1995) 'A European Perspective on the Spatial Links Between Land Use, Development and Transport' in D. Banister (ed) *Transport and Urban Development*, E&FN Spon, London, pp65–84

Hillman, M. (1999) 'The Impact of Transport Policy on Children's Development', *Canterbury Safe Routes to Schools* Project Seminar, 29 May 1999

Himanen V., Nijkamp, P. and Padjen, J. (1993) 'Transport Efficient Mobility, Spatial Accessibility and Environmental Sustainable Mobility' in P. Nijkamp (ed) *Europe on the Move: Recent Developments in European Communications and Transport Activity Research*, Avebury Gower, Aldershot

Kinnersly, P. (2001) 'I Quit', *World Transport Policy and Practice*, vol 7, no 1, p5–7.

Knoflacher, H. (1997) *Landschaft ohne Autobahnen: Für eine zukunftsorientierte Verkehrsplanung*, Böhlau, Vienna

Ostrom, E. (1990) *Governing the Commons: the Evolution of Institutions for Collective Action*, Cambridge University Press, Cambridge

SPD (2007) (Sozialdemokratische Partei Deutschlands) 'Hamburger Programm – Das Grundsatzprogramm der SPD', http://parteitag.spd.de/servlet/PB/menu/1727812/index.html, accessed 4 April 2008

T&E (2003) (Transport and Sustainable Mobility) 'The Social Pillar', *Fact Sheet*, August Brussels

T&E (2007) (Transport and Sustainable Mobility) 'Market based instruments (MBIs) in EU Transport and Environment Policy', *Position Paper*, July, Brussels

Wegener, M. (1995) 'Accessibility and Development Impacts' in D. Banister (ed) *Transport and Urban Development*, E&FN Spon, London, pp157–161

Wynne, B. and Jasanof, S. (1998) 'Science and Decision-Making' in S. Rayner and E. Malone (eds), *Human Choice and Climate Change, Vol. 1: the Societal Framework*, Batelle Press, Columbus, OH

4

Ageing

Introduction

Pension reform has proved to be a persistent policy issue. Throughout the 1990s, policy-makers in many countries have tinkered with old-age pensions. In Europe, the stage of many but by no means all of the reforms, policy-makers in some countries, particularly in central and eastern Europe, have radically reformed their old systems; others, typically those in affluent continental Europe, have been far more cautious, opting instead for incremental changes to existing pension systems. Others still have managed to implement fairly broad changes without making pension reform an explicit policy issue. In many parts of the developing world, particularly in the emergent economies of Asia and Latin America, policy-makers have also experimented with a wide range of pension schemes (Gillion et al, 2000; Gu, 2001; Liu et al, 2006).

Not only has pension reform been a persistent feature across different political systems, it has also shown incredible resistance to resolution over time. Just as policy-makers seemingly despatch the problem, pension reform worms its way back onto the policy agenda giving rise to a new round of what is, more often than not, a politically divisive and conflict-ridden policy debate. No matter whether the particular country reforming old-age pensions is affluent or in transition, corporatist or market-oriented, developed, emergent or developing country, pension reform is difficult.

Why is pension reform so difficult? Part of the reason is that it has become the focus of policy debate about demographic ageing. The next section shows that population ageing, despite the hard data that demographers provide, is a highly uncertain and complex issue area. The following sections, then, try to make sense of the policy debate about pension reform by analysing its scope, structure and potential impact.

The challenge of demographic ageing

Across the globe, populations are ageing. For many, this is the single most important challenge for societies today (World Bank, 1994; OECD, 1998;

European Commission, 1999). Populations in the developed world have aged because of two concurrent developments. First, due to improvements in medical care, nutrition and working conditions, people are living far longer than their parents and their grand-parents. Since 1960, the average male life-expectancy in the countries of the Organisation for Economic Cooperation and Development (OECD) has increased by eight years from 68 to 76 (OECD, 1998, p10). Second, for less apparent reasons such as increased female education, fertility rates in the north declined dramatically about 40 years ago. Since then, fertility rates in the OECD have declined by an average of 2.4 child per woman (Sleebos, 2003, p13). While total fertility rates in the OECD vary, only four countries could boast a fertility rate above the so-called replacement level of 2.1 children per woman in 2004 (OECD, 2006).[1]

In the developing world, populations are ageing somewhat differently. While fertility rates in the developing world have declined precipitously, increases in life expectancy have considerably lagged behind those in the developed world. Between the 1970s and the 1990s, median total fertility in the developing world declined from 5.9 to 3.9 children per woman (UN Pop, 2003, p.xvii). However, for reasons that the next chapter explores in more detail, average life expectancy in the poorest regions of the world is about 25–27 years below that of the richest countries (WHO, 2007). As a result, demographic ageing in the developing world is more rapid and intensive (UN Pop, 2007, p.xxvii).

Demographic ageing is likely to have significant impacts on our societies, both rich and poor. An area that has arguably attracted most attention is the probable impact of ageing on labour markets. Starting about 2010, the large age cohort, commonly referred to as the baby-boom generation, will begin to retire from the labour market in much of the developed world (World Bank, 1994; OECD, 1998; MacKellar, 2000). Thus, labour market growth rates are set not only to decline but turn negative around about 2010 (OECD, 1998; MacKellar, 2000). The European Commission (1999) has calculated that the share of the younger cohorts in the population will drastically fall. By 2015, the cohort of 0–14 year-olds will fall from 17.6 per cent of the population (1995) to only 15.7 per cent. The decline in the 15–29 year-olds, that is those that feed the labour market, will be about 16 per cent (European Commission, 1999, pp7–8). Contrast this to the rapid growth of older cohorts. Starting in 2010, the 60–64 age cohorts will grow by 26 per cent which corresponds to 16 million people across the European Union. The predicted growth rates of the aged and very old are particularly impressive: the 65+ age group will increase by 30 per cent and the 80+ group will grow by 40 per cent (European Commission, 1999, p8). At present, the age-dependency ratio[2] in the OECD varies from about 55 per cent (in South Korea) to 80 per cent in Mexico and Turkey (OECD, 2006, p42). This, however, is set to increase steeply: by 2050, age-dependency ratios in countries such as Japan, Italy and Spain are projected to increase by over the 100 per cent mark (OECD, 2006, p43).

If productivity gains do not rise dramatically from their post-1973 annual average of 1.5 per cent – and there are no compelling reasons to think that they will – then higher dependency ratios and lower labour market participation mean reduced economic output and a loss of material living standards (OECD, 1998, p10). What is more, if we assume that real disposable income affects fertility decisions, then, other things being equal, falling real incomes are likely to depress fertility rates even further.

However, while it stands to reason that demographic ageing will affect labour markets and the economy in some way, considerable uncertainties surround the exact nature of these impacts. Take for example, the role of migration in the future composition of populations. Brian Abel-Smith (1993) contends that the developed countries' unfavourable age structure could be balanced by an influx of tax-paying immigrants. While this is true in a purely arithmetic sense, the direction of this variable is difficult to predict. On the one hand, streams of immigrants from developing countries to industrialized nations have been steadily increasing in the past two decades. On the other hand, immigration policies in the industrialized countries have become increasingly restrictive. The German economists Börsch-Supan and Miegel (1999) contend that, at least for Germany, the level of immigration needed to offset the effects of ageing in the labour supply are likely to prove politically contentious. The net effect of these two countervailing forces is hard to foresee (Abel-Smith, 1993).

Further, the impacts of ageing will depend on the overall state and developmental path of different economies. Here, precise long-run economic predictions needed for policy-making are also difficult to obtain. Abel-Smith observes that the pension reform debate has fed on the decline of economic growth and the institutionalization of structural unemployment that has been evident in industrialized countries since the early 1970s: 'It was only when economic growth faltered that people began to write about the 'crisis' in the welfare and a period of modest retrenchment began' (Abel-Smith, 1993, p265). A return to a tight labour market, he continues, would appease the present climate of crisis. Others claim that a tight labour market may exacerbate the impacts of ageing due to skill and training mismatches. Given existing rigidities in the educational sectors on the one hand and the highly skilled nature of most jobs in OECD countries today on the other, the disappearance of a whole cohort of highly trained workers will leave firms floundering. Yet, whether labour markets in Europe will return to full employment, and under what conditions, is difficult to foresee. Global markets are becoming increasingly interdependent which means that domestic economic performance may have little to do with domestic economic policy. The more complex and interdependent the markets become, the higher the uncertainties of key economic variables: essentially, Abel-Smith contends, expert projections '... amount to no more than possible future scenarios. They are no more robust than the assumptions underlying them' (Abel-Smith, 1993, p260).

Lower fertility rates and higher life expectancy are also likely to affect social structures. For example, household sizes and household compositions will change considerably. Projections for the USA estimate that, by 2050, the share of single-parent households will increase from 8.5 per cent in 2000 to 10.9 per cent (Jiang and O'Neill, 2006, p30). Households consisting of couples with one child will fall from 31.7 per cent in 2000 to 27 per cent in 2050. Single-person households, in turn, are set to increase by just under three percentage points from 25.8 per cent to 28.4 per cent in 2050 (Jiang and O'Neill, 2006, p30). The share of homes hosting three generations, already small, is likely to dwindle to 1.8 per cent.

These trends will impinge on a wide range of social institutions in a myriad of ways. The family will come under significant adaptive pressure. At a time when the demand for care is likely to increase, family support structures will be diminished by demographic ageing, and stretched by mobility (both spatial and social). Additionally, growing labour market participation of women, who carry the main burden of caring for both the young and the old, will exacerbate informal family care arrangements.

The extent to which formal arrangements for care can replace family care is a wide-open question. Prohibitive costs of residential care and growing disaffection for institutionalized care mean that nursing homes are unlikely to make up for the shortfall. But providing care to one- or two-person households will require considerable innovation in social and health care service provision (Walker, 2002). Most importantly, new forms of mobile care will require human resources in labour markets that are experiencing shortages. It is not at all clear where these professional carers will come from. Again, immigration may be one, albeit increasingly contentious, option.

Alternatively, the ageing population itself, particularly the so-called 'young old' generation, may be a source of care. This, however, depends on how well the baby-boomer generation itself ages, and on the limits of longevity. Unfortunately, there is no certainty about either of these factors. While some biomedical researchers in the USA, particularly the US Pension Administration, believe that there is a practical ceiling to human longevity at about 85, others are predicting that a child born today may reach the age of 95 or even 100 with no theoretical upper limit (Roush, 1996, p4).

But all of this, of course, assumes that the demand for care increases with demographic ageing. This need not be the case, argue some experts. It is commonly assumed that health-care needs increase over the age of 80. Yet, Abel-Smith points to a study that shows that 'contrary to common belief, the costs of those who die aged 80 or over are about 80 per cent of the costs for those who die aged 65 to 79' (Abel-Smith, 1993, p261). The OECD Bulletin reports, however, that before the age of 80, the health differences across different age groups are not so marked. In fact, they argue that the health differences within a particular age group are '... usually more marked than across different age groups' (Hicks, 1997, p19). Add to this the fact that life-style changes will affect the health of the future aged and it becomes clear that these variables are very difficult to predict with any certainty.

The changing demographic composition of societies may also reconfigure prevalent patterns of political interests. Many fear that the increase in the age of the so-called 'median voter' will narrowly focus political attention on single-issues of interest to older voters (Metz, 2002). This, they continue, would come at a time when the fiscal room for manoeuvre will be severely limited. Thus, demographic ageing could divert funds from other, more important future-oriented projects, such as research into environmental technology or development of innovative transport infrastructure. This scenario, however, depends on the ability of older people to politically mobilize around the notion of age. Nothing so far, however, indicates that age will replace more traditional motives of political mobilisation, such as religion or political ideology. Indeed, age-based pressure groups and political parties for pensioners have had very modest success in shaping policy (Walker, 2002).

In most countries, pension reform has provided the platform for the public debate about the economic, social and political impacts of demographic ageing. This is because pension systems embody and reflect the economic, social and political contexts of ageing in any society. Despite the many different forms into which pension systems have evolved over time, they all transfer income from one generation to the next. A disproportionate increase in older, often inactive, people relative to younger workers, then, raises questions of how these transfers will be paid for and how this affects the economy. Further, pension systems are deeply embedded in – indeed are often an integral part of – the values and practices of a particular society. The rules and norms that regulate pension payments – who is to get what and why – are based on very specific perceptions of what constitutes a family (such as the male-breadwinner model), what are socially desirable activities (such as military service or child-rearing) or what makes up legitimate income inequalities. In this way, pension systems signpost,

via the retirement age, the boundary between work and rest, economic productivity and private consumption, the labour market and private sphere
(Esping-Andersen, 1990). The specific pensions mix in any one country,
that is the mix of sources of old-age income, reflects societal values concerning the tensions between efficiency and social justice. Methods of
financing pensions (see Box 4.1) institutionalize and reproduce intergenerational contract. Finally, pension systems also represent systems of
governance. The management of pension systems empowers certain policy actors, such as trade unions or private insurers. What is more, by
creating winners and losers, pension systems generate political preferences
for or against the status quo.

Box 4.1 *Methods of financing pension systems*

Pension systems are financed using a wide range of different institutional
mechanisms. In general, research categorizes these mechanisms as follows:

- Pay-as-you-go systems

The first type, the publicly managed pay-as-you-go system, is by far the
most common formal pension arrangement in Europe. Public pay-as-you-
go systems come in many different guises. In general, pay-as-you-go
systems define the level of benefits in advance (hence the term 'defined
benefit'). This implies that there is no actuarial relationship between contributions and benefits; this means that what an individual pays into the fund
need not have any actuarial relation to what they receive in terms of pension benefits. Yet, the precise way that systems define benefit varies widely.
Some systems (e.g. Australia) provide a flat, universal benefit regardless of
income or employment history. Others also provide a universal flat benefit but tie them to a certain number of contribution years (as in, for
example, the UK). Defined-benefit systems can also pay means-tested benefits or minimal pension guarantees. Yet other pay-as-you-go systems peg
benefits to earnings; this system, common in continental Europe, provides
higher benefits for those workers with previously higher incomes (World
Bank, 1994, p114).

Most often, contributions to defined-benefit public pay-as-you-go systems take the form of payroll taxes. Here, employees pay part of their
wages into the pension fund and employers contribute an equal part from
profits. Alternatively, policy-makers can partially fund public pay-as-you-go
systems from general revenues, thereby relieving the upward pressure on
unit labour costs.

The advantages of such a system are that it can easily, and fairly efficiently, redistribute pension income across different income classes. What

is more, a public pay-as-you-go system creates an intergenerational contract since current pension contributions represent future pension claims. In this way, younger generations are not only persuaded to forfeit consumption now for the prospect of consuming when they retire but also have a vested interest in the stability of the system. Additionally, public pay-as-you-go systems protect individuals from those risks relating to investment and market fluctuations as well as disability, longevity and individual risks. They remain, however, vulnerable to demographic risks and political risks.

- Traditional occupational pensions

The second type of formal system is the occupational pension. Here, individual firms or entire industrial sectors institute a pension fund for employees. Occupational pensions have the advantage that they involve relatively little administration cost. Moreover, firms can easily set up occupational pensions without much help (or, depending on your point of view, interference) from the public sector. Here, the World Bank argues that, in contrast to pay-as-you-go pensions contributions, workers will tend not to perceive contributions to occupational pensions as a tax.

In general, the private sector is responsible for managing occupational pensions. The particular management forms of occupational pensions vary widely. Occupational pensions are traditionally defined-benefit plans (although defined-contribution occupational schemes are becoming increasingly popular); they can be fully funded, partially funded or completely unfunded; occupational pensions can be tied to one particular firm or to an industrial sector. In any case, occupational pensions are subject to heavy regulation (World Bank, 1994; OECD, 1998; Gillion et al, 2000). Accordingly, the risks that these schemes are vulnerable to depend on the precise set-up of the plan.

- Mandatory and voluntary savings plans

Whereas the previous two types of pension systems are well established, the latter two are forms that are relatively new and not common. This type of pension scheme takes the form of occupational pensions (only if they are defined-contribution), personal saving plans and annuities. These can be either mandated by the government (as in many newly reformed Latin American countries, most notably Chile) or voluntary schemes where the government often offers financial incentives (such as preferential tax treatment as in the USA).

Essentially, products purchased from financial institutions, these plans provide individuals with a means of saving income for retirement. These defined-contribution plans are, by definition, fully funded, meaning that benefits relate directly to contributions plus any capital gains. Here, individuals bear the investment risk and the risk of volatile returns inherent in market operations. Although there is no reason why public sector institutions should not manage these types of plans, advocates of mandatory and voluntary savings plans argue that the private sector is best equipped to manage these plans. This would insulate pensions against the risk of political manipulation and the associated risk of imprudent investment (World Bank, 1994).

• Notional accounts

An alternative to funded defined-contribution savings accounts is the idea of notional defined-contribution accounts. Here, workers accumulate pensions contributions based on a contribution and a notional interest rate (which can be the market interest rate or, as in Sweden, an alternative indicator reflecting economic growth). At retirement, pension managers transform the account into annuitized benefits. The scheme, however, is not funded: no actual capital reserves back up the accounts, meaning that current contributions continue to finance current pensions.

The advantages of this approach, very recently introduced in Sweden, are to make the pension system more transparent by more closely relating benefits to contributions. Since the system is not funded in advance, the individual bears the longevity risks while society bears demographic risks and economic risks.

Pension systems, then, are extraordinarily sensitive to the way demographic ageing affects these economic, social and political contexts. That is why pension reform has become the focal point of the debate about demographic ageing. And since pensions are irredeemably entangled in the complexities and uncertainties of demographic ageing, debate about pension reform gives rise to intractable policy controversy. Demographic data provides us with a quantitative picture of the way age profiles in the developed and developing world are likely to progress. This data, however, tells us very little about what will it be like to live in a society in which the median age has increased by 10 years.[3] Yet, policy responses to demographic ageing need some idea, some vision of how society will evolve. Policy-makers, then, interpret demographic data and judge which of the many impacts and future scenarios is most plausible. In order to fill in the gaps, contending advocacy coalitions rely on their divergent frames to help them make sense of demographic ageing. Only then do these competing political constituencies get an idea of what to do about pension systems.

The scope of policy conflict about pension reform

Using the typology of frames, introduced in Chapter 2, the following section reconstructs three contending policy narratives about pension reform.
Like any good yarn, the policy stories about pension reform create settings
(the basic assumptions), villains (the policy problem and who or what is
causing them) and heroes (the policy solution and who or what should be
responsible). Similar to transport policy, each contribution tells a slightly
different story of the same issue: each identifies problems, apportions
blame and claims to provide solutions. Each story combines factual obser-
vation with fundamental beliefs about how to best manage pension
systems. In short, each policy story frames the pension issue in a particular and conflicting way.

Crisis and intergenerational equity

The first story throws down the gauntlet to established pension systems.
While the world has changed – not only in demographic terms – pension
systems have remained pretty much the same since they were invented in
the 19th century. For advocates of this discourse, including the World
Bank, the financial and insurance industry as well as a host of (mostly
libertarian) think-tanks such as the Deutsche Bank Research, the Cato
Institute and the Adam Smith Institute, pension systems are steam-
powered machines with shiny brass gauges and wood panelling trying to
regulate a world of six billion people run by broadband internet and nanotechnology. More than that, the inflexibility and expense of these grand
old machines has become a millstone around the neck of contemporary
societies. Pension systems are, proponents argue, no longer up to the job,
and it is time we allowed them to retire.

The setting: a world of scarcity and tough choices
On this view – the crisis story – the world is a place of fundamental
scarcity. More of one thing invariably means less of something else. In a
world of scarcity and trade-offs, then, all the policy-makers need do is to
provide the framework in which rational economic actors can properly
identify and assess these trade-offs. Understanding trade-offs and scarcity,
proponents of this policy story contend, becomes all the more important
in the context of increasingly harsh competition which leaves little room
for mistakes. Only firms and, by extension, economies that are competitive
will be able to succeed in the new global economy.

For proponents of the crisis story, pension systems are little more than
tools for transferring income from one group to another. In particular, this

policy story argues, pension systems are supposed to do two fundamental things. First, pension systems must provide an adequate level of income in old age. Second, pension systems must also promote (or at least not hinder) economic growth. Everyone, the World Bank (1994) reminds us, '... old and young, depends on the current output of the economy, so everybody is better off when the economy is growing ... and in trouble when it's not' (p3).

On this view, then, policy-makers ought to assess pension systems in terms of how efficiently they provide old age income and promote economic growth – no more, no less. While social justice (gender equality, intergenerational solidarity, redistribution, etc.) is without a doubt a worthy aspiration, pension systems are poorly equipped for achieving this type of goal. If policy-makers want fairness, so the argument goes, they are well advised to concentrate their efforts on things that weaken the inherent, if at times rough, fairness of competitive market.

The villains: inefficiency and inequity

The main problem with pension systems, argue proponents of the crisis story, is that they are no longer efficiently doing the jobs for which they were designed. Today, so the crisis story argues, pension systems are being squeezed from three sides. First, demographic ageing is rapidly undermining the financial foundations of pension systems. Demographic ageing, so the argument goes, will significantly and permanently change the age structure of societies around the world. For pension systems that finance current pensions from the contributions of current workers – so-called pay-as-you-go systems – demographic ageing means that an ever-decreasing base of contributors has to maintain an ever-increasing volume of pensioners. Add to this the fact that this process will take place relatively rapidly as the large baby-boom cohorts start to retire from about 2010 onward and you have a financial crisis of the pension system.

Second, argue proponents of the crisis story, the demographic crunch will set in at a time when economies are under severe competitive pressure. If economies are to survive in these harsh new conditions, so advocates of the crisis story argue, firms will have to keep production costs low. Demographic ageing, however, will send social insurance costs through the roof. Businesses in countries with an already heavy demographic burden, such as most European countries, will simply be unable to compete. Structural unemployment, a further drain on the contribution base of social insurance systems, is the inevitable consequence. Moreover, economic globalization, the transformation from a mostly industrial to a mostly service-based economy as well as rapid progress in production technologies will additionally squeeze the contribution base.

Third, to add insult to injury, inherent design flaws of pension systems themselves magnify the mounting financial pressures on public pension provision. At a time when global competition is exceptionally harsh, pension systems distort labour markets, depress national savings and detract public funds from urgent social projects elsewhere (e.g. education). In most European countries, so proponents of the crisis story argue, pension systems have enabled workers to retire before the statutory retirement age. While this helped ease tense labour markets in the short-run, it has also created a mountain of so-called implicit debt to be serviced by the younger generation of workers (Brooks and James, 1999). What is more, proponents of the crisis story argue that public pension systems divert much-needed resources away from financial markets. Just when societies, particularly in Europe, should be investing in the future – by funding education, research and innovation – they will be forced to service a crushing pension debt. This is why the German Council of Economic Experts warned that the ...

> ... *current old-age pension system constitutes a problem for the economy as a whole, as it has negative impacts on other sectors: high contribution to the old-age pension scheme creates undesirable incentives, for example, on the labour market, where, because of the high level of social security contributions, workers move on to jobs in which such contributions are not obligatory. (p13)*

The three squeezes, argue the proponents of the crisis story, spell nothing less than financial and economic disaster for European pension systems. Sky-rocketing social costs, caused by demographic ageing but also by poor policy decisions of the past, will hobble European firms as they try to compete in increasingly global markets. What is more, current pension policy is grossly inequitable to younger generations of workers who, through no fault of their own, are left to foot a huge bill of implicit pension debt (World Bank, 1994). As their contribution rates increase to something like half of their income, the rate of return they can expect from this investment is scandalously small, even negative. Techniques such as inter-generational accounting show that present pensions policy amounts to a swindle of enormous proportions with younger workers the butt of the scam.

Why, then, do policy-makers not rectify what is, quite obviously, screaming injustice? The simple but sad truth, say proponents of the crisis discourse, is that policy-makers, self-serving and driven by the vote motive (Tullock, 1976), will not endanger their chances of re-election by inflicting losses, however justified by economic rationality, on what is swiftly becoming an immensely influential group among European voters (Pierson, 1994; Bergheim et al, 2003). A powerful but well-organized minority, the so-called 'pro-welfare coalition', catalyzes and concentrates the forces in favour of the hugely inequitable status quo. This coalition of actors, composed of trade unions, social NGOs, Old-Labour-style politicians, and social insurance bureaucrats, successfully and consistently mobilizes the 'grey vote' by driving the fear of the market into an already risk-averse European electorate (Pierson, 1996; Bergheim et al, 2003).

The heroes: playing to institutional strengths

This is a tragedy, maintain the proponents of the crisis story, because the solutions to this impending financial and economic apocalypse are to hand, if only someone would show political will and take the initiative. Policy-makers need to redesign pension systems so that they can withstand the mounting pressures of demographic ageing and economic globalization. This, they contend, involves allowing different types of institutions to play to their strengths in old-age income provision. At present, huge monolithic pension systems in the public sector are responsible for the whole range of tasks associated with old-age income: long-term savings and securing living-standard in retirement on the one hand and redistribution and coinsurance for associated social risks (disability and longevity) on the other. Rather than the public sector spreading itself thinly and doing all jobs badly, the proponents of the crisis story suggest that policy-makers should let it concentrate on those tasks it is good at (redistribution and coinsurance) and let the private sector take over the rest (long-term savings and wealth creation) (World Bank, 1994; Börsch-Supan, 1999).

Specifically, this involves separating pension systems into functional institutional pillars (World Bank, 1994). The first pillar, run by the public sector, would concentrate solely on the prevention and amelioration of old-age poverty. Here, modest but adequate flat-rate universal benefits would prevent destitution in old age while not acting as a disincentive to work. Instead of levying crippling and unpopular pay-roll taxes, this pillar would draw its funds from general revenues. This would also provide policy-makers with far more leeway to redistribute from lifetime rich to lifetime poor since they would not be redistributing contributions. The second pillar, responsible for long-term savings, would allow individuals to smooth their income over the lifecycle. Rather than a monolithic public sector pillar, advocates of the crisis story envisage a vibrant market of private sector actors offering a wide selection of financial products. However, since people are inherently myopic, this pillar requires a degree of compulsion in order to prevent free-riding and old-age poverty (World Bank, 1994): workers are obliged to invest a certain percentage of the net wage (say 10 per cent as in Chile or Poland) into a pension fund of their choice. Third, for those who want and can afford further protection, policy-makers should institute and promote a third pillar consisting of voluntary savings of any sort.

Multi-pillar systems, so the argument goes, are far more resilient to the pressures of demographic ageing and economic globalization for several reasons. First, by lowering unit labour costs, European firms will be able to compete more effectively in global markets; this will increase economic growth which, as we have seen, makes everyone better off. Second, shifting to a fully funded, defined-contribution pension scheme also introduces more transparency and fairness into labour supply decisions. Actuarially neutral pension systems do not displace the costs of early retirement onto the society, specifically the younger generation of workers. Removing these labour market distortions, proponents of the crisis story contend, will invariably lead to more informed and more rational retirement choices. Third, flushing the enormous volume of funds from the public sector pension systems through the private capital markets, proponents of the crisis story maintain, will boost European financial markets, again leading to more economic growth. Last, by extricating old-age income from the meddlesome influence of self-seeking politicians, multi-pillar systems will immunize pension systems to political abuse.

Social stability and solidarity

The second policy story – the stability story – in the European pension reform debate beseeches policy actors to remain calm. It is a rather defensive tale told by the much-maligned 'pro-welfare coalition' (Bonoli, 2000; Bonoli and Palier, 2001). Trade unions (such as the ILO – International Labour Organisation – at international level, or the DGB at national level), members of the welfare state bureaucracy (such as the ISSA at international level, and the *österreichische Sozialversicherungsanstalt* at national level), as well as Old-Labour and Catholic Conservative politicians tell a rueful story of the way an institution at the very heart of European societies is being dismantled. Europeans, misled by the financial industry and abandoned by their politicians, have engaged in a perilous and risky experiment with potentially irreversible consequences for social peace and stability in Europe.

The setting: pensions ensure stability

The important thing to remember about pension systems, proponents of the stability policy story tell us, is that they have been an unmitigated success wherever they have been allowed to flourish. On this view, pension systems are more than mere instruments for transferring income (Gillion et al, 2000; ILO, 2007). Above and beyond the profanely pecuniary, pension systems institutionalize social solidarity. Pension systems embody a commitment to social peace and stability.

They do so by suspending the contradictions between capitalism and citizenship (Marshall, 1950). Pension systems represent the 'irenic formula' that has successfully married social progress with capitalism thereby bringing to an end the divisive social conflicts that have torn European countries apart in the past (Hinrichs, 1998). Not only have pension systems kept the lid on social unrest by eradicating old-age poverty, they have also provided European business with stable industrial relations as well as a highly skilled and motivated workforce (Hinrichs, 1998; SoVD, 2000). The unprecedented increases in the European standard of living in the past four decades '... can be attributed to the creation of social security pensions which must be considered as one of the greatest social developments of the last hundred years' (Gillion et al, 2000, p1).

Successful pension systems, particularly in Europe, have been an integral part of nation-state development. The ups and downs of social insurance systems, proponents of this story remind us, have mirrored the turbulent history of European nations (VDR, 2000): over the past hundred years or so, European social insurance systems have survived two world wars, hyper-inflation, as well as the post-war boom and bust periods. On a continent with such a troubled history as Europe, institutions that can guarantee stability and social peace over time are not to be tampered with light-heartedly; certainly not by amateurs out to make a quick buck.

By the same token, proponents of the stability story point out that countries without the benefits of strong pension systems and welfare states, are riven by social division and strife. In most of the developing world but also in countries like the USA, inadequate social protection of workers, including retired workers, exacerbates the huge observed disparities in income, education, health and well-being. For proponents of the stability story, socio-economic disparities rapidly turn into social unrest and disorder. It is no coincidence, argue proponents of the stability story, that a rich country like the USA has appalling crime rates compared to wealthy countries in Europe.

This is why the International Labour Organisation (ILO, 2001) is convinced that social security is ...

> ... *very important for the well-being of workers, their families and the entire community. It is a basic human right and a fundamental means for creating social cohesion, thereby helping to ensure social peace and social inclusion. It is an indispensable part of government*

*social policy and an important tool to prevent and alleviate poverty.
It can, through national solidarity and fair burden sharing, contribute to human dignity, equity, and social justice. It is also
important for political inclusion, empowerment and the development
of democracy. (ILO, 2001, pp1–2)*

The villains: scaremongering

The real problem with social security is, quite simply, that there is not
nearly enough of it. Most workers in the world, the ILO (Gillion et al,
2000; ILO, 2001) points out, are not covered or are covered inadequately
by any form of social insurance or pension scheme. And for the workers
in the developed world who have fought for adequate social security, political forces serving narrow employer interests are now dismantling these
foundations of social stability and peace.

In most of the developing world, workers in the informal sector, in agri-
culture, household workers (mostly women) and the self-employed have
no entitlements to retirement income. Coverage, the ILO argues, usually
depends on several factors. These include the level of economic development and the age of the pension systems: the richer the country and the
older the pension system, the higher the coverage is likely to be. Further,
the way in which governments finance pensions as well as the capacity of
the social security administration have an impact on coverage: defining the
contributions base as well as enforcing contributions payments significantly affects the level of coverage that policy-makers can realistically
achieve. Last, the ILO argues that government policy – that is to what
extent pensions are a policy priority – will significantly affect coverage
(Gillion et al, 2000, p8).

Another major pension issue in developing countries, the ILO contends,
is the governance of pension schemes. In the past, public and private
management of pensions, at least for the vast majority of workers in the
developing world, has been poor. The reasons for bad pensions management are manifold. Often pension systems have been politicized and abused
to achieve short-term political aims. Many pension systems suffer from
inherent design faults and badly conceived administrative procedures. In
general, this has resulted in high administrative costs coupled with poor
services (Gillion et al, 2000, pp8–9). The implication here is that poor
management, rather than inherent pension scheme design, makes
pension unattractive to workers who then withdraw to the informal
sector.

What about pension systems in the affluent North? What about the
impending financial catastrophe brought about by the triple squeeze of
demographic ageing, economic globalization and inherent design faults?

In the developed world, advocates of the stability story take a very dim view of what they consider to be scaremongering by interests for whom a crisis in social security would be politically convenient, not to mention inordinately lucrative. Demographic ageing, unemployment and globalization, it is true, strain the functioning pension system. Additionally, something which the crisis story likes to ignore is that labour market flexibilization has meant that typical employment relations are on the decline. Whereas social insurance systems assume continuous employment histories at or above the average wage, these traditional career patterns are rapidly becoming the exception rather than the rule.

But, ask the proponents of the stability story, does this constitute a crisis? A crisis, no less, that requires us to jettison tried-and-tested pension systems for risky experiments with the free market? Certainly not. Like all parameter changes, these developments and phenomena are neither new nor are they particularly dramatic. They are in fact purely operational issues. Dealing with them effectively and competently is all in a day's work for the dedicated and highly trained professionals who captain these social insurance systems. These challenges alone certainly do not account for the crisis so vociferously invoked by policy actors with a vested interest in undermining trust in and the reputation of the public pension system.

Rather, so the proponents of the stability story contend, the fundamental issue is that workers in the developed world, particularly in Europe, seem to have turned their backs on social solidarity. Instead, people in all walks of life and at all levels of society seem to have adopted the ruinous ideology of 'individual responsibility' (Rehfeld, 2001). This shift is apparent at all levels of society including, unfortunately, the level that counts most: the political elite (Seeleib-Kaiser, 2002). Partly due to a generational change within political elites (witness the shift from Old to New Labour or from One-Nation Tories to Thatcherism), but also partly due to the global spread of the neoliberal Washington Consensus, welfare states and the principles of social solidarity they embody have become highly unfashionable.

The consequences of this shift in societal values, contend advocates of the stability story, are likely to be severe for stability and social peace. First, this orientation towards 'individual responsibility' has led to a fetish-like obsession with the costs of social security. Pension reform debates focus exclusively on the allegedly intolerable levels of social insurance contributions. This, contend advocates of stability, has driven governments to retrench pension benefits for marginal workers, particularly the disabled, beyond what would be ethically or even economically justified (Schmähl, 2000). For example, after the barrage of pension reforms in Germany throughout the 1990s, workers now need work for 26 years at the average wage to receive a pension at the level of social assistance (Hinrichs, 1999;

Schmähl, 2000). Similarly, a working-life's worth of National Insurance payments in the UK is not nearly enough to avoid severe poverty and deprivation after retirement (Hill, 2007).

Second, by blaming public pension systems for problems that are beyond the control of social insurance institutions (such as economic globalization or labour market policies that have let tens of thousands of industrial jobs go to the wall), the debate has stylized private sector, fully funded pension systems as the panacea to all of society's ills. This, argue the advocates of the stability story, is based more on wishful thinking than hard-nosed facts. Not only do these mandatory fully funded, defined-contribution schemes distort markets and (see Box 4.2), they actually offer workers less protection at higher costs. So, in addition to distorting labour markets and encouraging fraudulent pension governance, these schemes are incapable of creating predictable pension payments, cannot maintain their value over time (since indexing private pensions is costly) and provide no means of democratic control (Gillion et al, 2000, pp22–23). Far from relieving the financial pressures on younger workers, a shift to such a fully funded, defined-contribution system would impose an intolerable double burden on the intermediary generation who would have to finance both accrued pension rights as well as save for their own retirement. Advocating reforms of this type, so proponents of the stability story tell us, seems sheer madness.

Box 4.2 *Problems with fully funded, defined contribution pension systems*

The proponents of the crisis story, advocates of stability lament, have painted an all-too-optimistic picture of fully funded, defined-contribution systems. However, the alleged superiority of fully funded schemes over traditional pay-as-you-go schemes does not stand up to scrutiny.

First, Gillion et al (2000) argue that the problems commonly associated with public defined-benefit, pay-as-you-go systems apply equally to private mandatory defined-contribution saving plans. Like any mandatory scheme, private fully funded, defined-contribution schemes seek to change the behaviour of rational economic agents by coercion. Any type of mandatory scheme '... will cause distortions, as individuals act to minimize the consequences of the programme that is undesired by them' (Gillion et al, 2000, p12). Further, if individuals are risk-averse, then private sector insurers will, if they are to attract customers, provide features that reduce risks to pensioners. However, these features, argues the ILO, break the link between contributions and benefits creating all the distortions usually ascribed to public defined-benefit pay-as-you-go plans (Gillion et al, 2000, p13). Indeed, many existing defined-contribution schemes contain these elements which include guaranteed minimum benefits, rate-of-return guarantees, or benefits based on rates of return fixed by pension funds. Add to that the fact that private defined-contribution plans tend to have higher administration costs and are very vulnerable to capital market risks, and the World Bank reform proposals do not look as attractive. What is more, if management is left to the private sector, the opportunities for malfeasance, mismanagement and private sector incompetence are, the ILO argues, practically limitless. In short, the ILO argues that placing responsibility ...

> ... for managing the considerable sums of money in mandatory defined-contribution pension accounts in the hands of private pension fund managers requires some mechanism to ensure that those funds are not stolen or otherwise misused. (Gillion et al, 2000, p7)

If individuals are responsible for managing their retirement income, then states have to ensure that workers are properly informed. Finally, if the state is in charge of managing mandatory defined-contribution savings plans, then policy actors should avoid the politicization of pension funds. In either case, the ILO avers, capital markets will require careful regulation and governance by experts. The upshot of the argument is that there are no inherent benefits in fully funded, defined-contribution systems that could not be accrued by careful management and design of pay-as-you-go defined-benefit systems.

Second, advocates of stability argue that a switch to a fully funded scheme would not relieve problems for policy-makers. Would-be reformers, so the argument goes, seem to think that funded defined-contribution schemes provide a once-and-for-all solution to the fundamental problem of population ageing: the fact that more retirees depend on a smaller number of workers. In order to relieve the financial burdens of population ageing, the ILO continues, defined-contribution schemes will have to either reduce the benefits relative to income from work or increase the retirement age (or both). Yet, these are precisely the policy options open to pay-as-you-go defined-benefit systems: the fundamental problem of having to transfer more income across generations remains. In pay-as-you-go systems, policy-makers increase contributions to increase transfers; in funded defined-contribution schemes, pensioners must sell their assets to the working population. For a given level of benefits, the amounts that the young transfer to the old is the same. Thus, the ILO concludes, whether a pension system is funded or unfunded makes little difference to the way it has to react to population ageing (either increasing contributions or reducing benefits). Stated differently, funded defined-contribution schemes face exactly the same problems as unfunded defined-benefit plans (Gillion et al, 2000, pp22–23).

Sadly, argue the advocates of stability, there is plenty of method to this madness. Throughout the 1990s, the financial and insurance industry has made a concerted effort to undermine the public's trust in social insurance institutions. These interests have not shied away from creating insecurity and fear among the insured. A politically motivated, shallow and essentially one-sided interpretation of globalization has given rise to scenarios of crisis and catastrophe; in fact, these interpretations are merely a smokescreen for cutting costs to employers and firms. In pitching their, inferior pension products, banks and insurances have been dishonest about the real impacts of planned reforms. For example, internal rate-of-return calculations conveniently forget to mention that they assume full-time, continuous working histories, something that is rapidly becoming a thing of the past. Or when arguing for cuts in contribution rates of public pension systems, advocates have been less than forthcoming about the downstream impacts on future pension levels. Throughout the 1990s, then, employer interests have done their utmost to undermine the ideological foundations of the pension system which has amounted to no less than the 'defamation of the welfare state' (SoVD, 2000, p14).

The heroes: if it ain't broke, don't fix it

In both the developed and the developing world, systems of social security need to adapt to new demographic and economic realities. In the developing world this means expanding the coverage of pension schemes as well as ensuring that they are properly governed. In the developed world, this means finding sound and balanced policy responses to the demographic, economic and social challenges. This, however, is a far cry from jettisoning the tried-and-tested pension schemes in favour of risky and unproven private sector pension funds.

There is, argue the proponents of the stability discourse, no use in hiding the costs of demographic ageing or pretending that there is a wonder cure to make everyone better off. Demographic ageing, globalization and structural economic change will inevitably incur costs. This is true in the developed as well as the developing world. The aim of any fair and balanced social policy must be to distribute these costs appropriately and judiciously across different social groups rather than lumbering the weakest with the brunt of the adaptation costs. Instead of embarking on a risky adventure in the global capital markets – a game that only the very rich can afford to play – policy-makers should use the repertoire of policy levers available to recalibrate and fine-tune the pension system. Rational management of demographic change within the institutional logic of existing welfare state structures is the only sensible and doable reform strategy.

What basic design principles, then, should guide this adaptation process? Given that different economies have very different characteristics and that these characteristics are subject to change, there can be, the ILO avers, '... no one universal perfect retirement income scheme' (Gillion et al, 2000, p16). Whatever the shape of the system, it must be able to alleviate poverty and provide low-risk retirement benefits.[4]

A multi-tiered system, proponents of the stability story contend, is the best way of securing these pension objectives. Each tier features specific risk and redistributive characteristics. The bottom tier is responsible for poverty alleviation: here, means-tested benefits, financed from general revenues, ensure that pensioners will not fall into poverty. The second tier, the ILO continues, consists of a pay-as-you-go defined-benefit pension system which provides secure and predictable retirement income relatively insulated from market risks. The third tier, a mandatory defined-contribution plan, provides security against demographic and political risks. The final tier, consisting of voluntary saving plans and other non-pension income, can provide extra income for those who can afford it without burdening pension systems (Gillion et al, 2000, p16).

This approach differs from the World Bank model in two important respects. First, the ILO model implies that policy-makers really cannot expect any benefits from leaving pension management solely to either the public or the private sector. Both management systems have in-built advantages as well as shortcomings: pension systems should be sufficiently flexible (or redundant) to be able to neutralize these weaknesses. Second, the four-tier model reveals the ILO's emphasis on the insurance function of pension systems. The four tiers provide a solid, predictable and stable bulwark against all possible risks of ageing.

Moving towards a multi-tiered system, the ILO argues, requires careful political management. The basis for any type of incisive reform is a basic political consensus. For this reason, the ILO urges policy-makers to consult the policy actors (in this case employers' and workers' representatives) at all stages of the process. Moreover, effective pension reform may involve a campaign of public education. Not only will the general public need to be told by experts about the pension issue, policy actors (such as parliamentarians) may also need to increase their level of knowledge and awareness about the impacts of ageing (Gillion et al, 2000, p17).

Social citizenship and basic security

The last policy story in European pension debates is an outsider. While the main contenders in the pension reform debate clash in the public sphere, the social citizenship story is a relatively quiet and marginalized voice. Its proponents – social policy NGOs of the so-called 'new social movements' (such as Age Concern but also traditional humanitarian organizations such as the Red Cross), by academics and by Green Party politicians – tell a story of how the discrimination and disadvantage of older people has made our societies less hospitable for all of us. Although marginal in the policy debate, this story is always present to point to the weaknesses of the other two approaches and outline the contours of an alternative.

The setting: emancipative social policy

The social citizenship tale urges us to take a broader and more inclusive approach to social policy in general and pension reform in particular. Social policy is not just about providing an adequate level of income, judiciously balancing burdens or even economic growth. Rather, the fundamental aim of any social policy is to guarantee self-determination and promote autonomy of all citizens regardless of age, gender, ethnicity, health or socio-economic background (Bündnis 90/Die Grünen, 2002). Social policy, so the argument goes, must empower individuals to take a full part in society by enabling them to claim their rightful share of socio-cultural, economic and political life. Social policy must enable individuals to fulfil their real human needs for socio-cultural inclusion and belonging. The German Green Party (Bündnis90/Die Grünen, 2002) argues that their ...

> ... notion of social justice and solidarity extends far beyond the classic policy of redistribution. Our primary policy goal is to prevent poverty and social marginalization and improve the social situation of those worst-off within society. We want to create stakeholder equitability, allowing everyone access to the key social areas of education and training, work and political participation ... In our view, the core issues in the question of equitable treatment deal with equality of opportunity between women and men, the rights for all the citizens of our country to participate equally in society, and, by the same token, the issue of equality between young and old across the generations. (Bündnis90/Die Grünen, 2002, p48)

This fundamental policy objective becomes all the more important in times of rapid social, economic, political and environmental change. Like the other policy stories, the proponents of social citizenship see the writing on the wall. Societies, they contend, stand at the brink of momentous social change: demographic ageing will transform all aspects of economic, political and social life. Such a comprehensive challenge calls for an equally thorough policy response. This, in turn, requires the application of a holistic lifecycle approach to the issue of ageing. Old-age income provision and social policy, argue proponents, is merely one arena in which these changes will unfold (Walker, 2002). Rather than concentrating on the costs of demographic ageing, the lifecycle approach relates individual and collective well-being over time to the complex interaction of a wide variety of factors (Walker and Naegele, 1999). These include family life, employment, education, socio-cultural participation, material security and health.

The villains: discrimination and exclusion

Current pension reform debates the world over, so advocates of the social citizenship position maintain, are missing the point. The basic issue here, argue proponents, is not whether to shift from a pay-as-you-go, defined-benefit to fully funded, defined-contribution or notional defined contribution or provident fund or what-have-you. These are technical details. The real question that policy-makers are steadfastly shirking is how best to deal with the profound socio-cultural transformations in store for us as our societies age. Societies, the proponents of the social citizenship story contend, stand at the brink of a momentous socio-cultural transformation (Walker and Naegele, 1999; Walker, 2002). Demographic ageing will sweep away many of the ideational and structural premises that are the foundations of traditional societies. Demographic ageing will throw out of kilter labour markets, health-care systems, as well as social protection systems. Most importantly, however, understanding and dealing with demographic ageing will require rethinking of many of our socio-cultural norms and practices. Demographic ageing is a unique opportunity, they argue, for much-needed socio-cultural renewal.

The main barriers to seizing the wonderful opportunities offered by demographic ageing for socio-cultural renewal, so the argument goes, are the pervasive and extensive inequalities that characterize societies around the globe. Wherever we look, even in affluent European economies, we find institutionalized inequalities between men and women, between rich and poor, between the indigenous and immigrant populations, as well as between old and young. Whether in the labour market, in health-care provision, in the social insurance system or in the mass media, contend the proponents of the social citizenship story, marginal social groups are systematically excluded from society. And, at global level, the vast marginal and poor regions of the world are excluded from the prosperity and wealth of the rich minority.

This is particularly (but not exclusively) true for older people. Rather than clumping together everyone over 65 in a homogenous group called 'the old', the proponents of the social citizenship story suggest adopting a more fine-grained view. For example, in Europe, the group over the age of 65 is highly heterogeneous in terms of income, social status, health, mobility and education. The popular image '... a new proletariat toiling long hours in order to pay taxes necessary to keep politically organized retirees in the style to which they have become accustomed' (Pirages, 1997, p40) is as pernicious as it is inaccurate. While it is true that some of the over-65-year-olds are affluent and politically influential, contend the advocates of social citizenship, very many are not. In fact, research shows that older people, far from being the sinister political force of Public Choice folklore, are poorly represented in political systems in Europe (Walker and Naegele, 1999). The same holds true for the developing world. In the labour market, older workers – particularly those with marginal skills, women and immigrants – are subject to systematic and pervasive discrimination (Walker, 2002). Vulnerable older workers are the first to feel competitive pressures and, once unemployed, find returning to the labour market virtually impossible. Discrimination of older workers, then, invariably leads to their exclusion from labour markets.

This is not only patently unjust, it is an almost immoral waste of resources. In a society where labour power will become increasingly scarce, argue the social citizenship contenders, it seems an awful waste to squander these resources.

The standard responses from policy actors in the pension debate, argue advocates of social citizenship, are woefully inadequate. In affluent countries, existing welfare states – particularly of the social insurance flavour – have done little to address these inequities. On the contrary, binding income-oriented pension benefits to participation in the formal labour market has merely transferred existing inequalities in the labour market into retirement. It is, the advocates of social citizenship point out, no coincidence that old-age poverty (which is far more prevalent than crisis advocates would have you believe) afflicts marginal social groups in the labour market: women, immigrants, the disabled, and low-skilled workers. However, market-based old-age income provision alone, contend the social citizenship advocates, is no alternative. Well known and well documented failures in the market for pension insurance mean that marginal groups are again left out in the rain. Moreover, private sector pension schemes will do nothing to empower and enable citizens; if anything at all, private sector pension funds will aggravate rather than alleviate social inequities.

In poor countries, the pension design debate is even more cynical. Existing social security and pension systems in the developing world more often than not benefit no one but a small and exclusionary elite. And, proponents contend, the market is mostly to blame for abject poverty leaving the old nothing but their (equally poor) families to rely on as they age. What the old in poor countries need, argue advocates of social citizenship, is rudimentary social and health services to meet their basic needs.

The heroes: active ageing and basic security

Since dominant pension policy actors at national and international level have reduced an issue of profound socio-cultural importance to little more than a tawdry question of bookkeeping, the advocates of the social citizenship story suggest that it is time to reclaim and expand the pension debate. Rather than focusing on questions of pension system design, important but not, in themselves, sufficient to address the problem of demographic ageing, we need to embark on a transversal and integrated strategy of 'active ageing' (European Commission, 1999; Walker, 2002).

Instead of merely coping with demographic ageing, proponents of the social citizenship story suggest actively shaping it. The fundamental premise of active ageing is that the best way of dealing with the costs of demographic ageing is to avoid them. That is why an integrated active ageing strategy addresses the issue in the key policy domains of social protection, the labour market and health-care systems. The modern welfare state, argue social citizenship advocates, should aim to provide a basic level of social protection based on real human needs (Bündnis 90/Die Grünen, 1998, 2002). Rather than creating unnecessarily bureaucratic and inequitable distinctions in welfare provision, the basic social security should be the same for everyone: old, young, immigrant, native, man, woman, healthy, disabled. This basic level of welfare will provide citizens with the material security to participate fully in social, cultural and political life. Unlike most current benefits, receipt of basic security would require no prior contribution. A first step towards this long-term basic income structure is establishing (where there are none, e.g. Germany) and strengthening (where they exist, i.e. France or Austria) basic pension provisions.

Apart from comprehensive social security, this basic universal income aims to ensure citizen autonomy. The basic, universal and unconditional income is supposed to emancipate citizens from dependency on either the state or the market. However, proponents of the social citizenship story point to the supply side of citizen autonomy. In addition to providing an unconditional income, then, comprehensive and emancipatory social policy needs to provide social services independently from central state control on the one hand and from commercial interests on the other. On this view, then, the production and provision of social welfare needs to be relocated into civil society; emancipatory social welfare for all ages, then, must rely on trusted NGOs rather than central bureaucracies or profiteering firms (Bündnis90/Die Grünen, 2002). This, proponents of social citizenship contend, is particularly pertinent for the old and vulnerable in poor countries where public institutions are ineffective and corrupt. Here, relocating the responsibility for social provision into the communities not only liberates the old and poor from the market and state strictures, it also helps build institutional capacity at local levels.

However, the proponents of social citizenship contend, a successful active ageing strategy will hinge on breaking down barriers and inequities in the labour market. Active ageing strategies will transform workplaces from sites of pervasive discrimination against marginal workers, whether they are old, the young, women, the disabled or immigrants, into open and inclusive social units (Walker, 2002). Starting with the physical make-up of workplaces unsuitable for older or disabled workers and ending with discriminatory hiring and promotion practices, an active ageing strategy addresses the problems of inequity at the root. This also includes providing the appropriate skills and training to allow workers of any generation to compete on a level playing field.

Since productive societal activity and participation presupposes good health, an integrated active ageing strategy places great emphasis on maintaining and promoting health (WHO, 2002). Again, prevention rather than amelioration is the key to success, contend the social citizenship advocates. By improving working conditions (by reducing stress levels at work), by educating citizens about avoiding health risks (such as smoking) and by fostering healthy practices (such as regular exercise) an integrated active ageing strategy could dampen the dreaded explosion of health-care costs in rich countries (OECD, 1998 – see next chapter).

Last but certainly not least, the advocates of social citizenship maintain, an active ageing strategy is predicated on a fundamental shift in socio-cultural perceptions regarding ageing and old age. The dominant paradigm of progressive physical degeneration and social disengagement will apply to an ever-smaller share of over-65-year-olds. In fact, argue the social citizenship advocates, these discriminatory perceptions of ageing, old age and older people will constrain us when facing the challenges and opportunities of demographic ageing. Most importantly, however, is that the emancipation of older people also addresses the societal inequalities of all other marginal and vulnerable social groups. Active ageing is not about 'grey representation'. Rather, active ageing, properly understood, is an emancipatory programme of socio-cultural renewal for everyone. In short, active ageing aims to build a fairer and more equitable society 'for all ages' (WHO, 2002; European Commission, 1999).

Table 4.1 displays the contending policy stories on ageing.

Table 4.1 Scope of policy conflict about ageing

	Crisis	Stability	Citizenship
Setting	Pension systems must promote economic growth; policy-making is about trade-offs and choice in a world of scarcity; deep suspicion of probity and competence of public sector	Social security institutions are more than technical implements for transferring income; social security brings about stability, order and predictability; promotes economic growth via healthy and highly skilled workforce	Adapting to an ageing society will require fundamental change at societal and individual level; pension reform will be one part of a broad strategy of socio-cultural adaptation; policy-makers need to take a life-course perspective rather than concentrating on specific age groups
Villains	Public PAYG, defined-benefit systems are inefficient and inequitable; public pension systems are no longer able to cope with economic, demographic and internal pressures; public pension systems distort labour markets, depress national savings and create sizeable intergenerational inequities	There is far too little social security to protect the world's workers; poor coverage and poor management characterizes pension systems in most countries; no panacea or technical fix to the fundamental issues of demographic ageing; financial industry has spread fear in order to undermine intergenerational solidarity	The real problem is pervasive discrimination of older people (along with other vulnerable groups); older people are systematically marginalized in contemporary society; narrowing the focus of ageing debate solely to pensions and the costs of ageing is short-sighted and unjust
Heroes	Multi-pillar systems to separate insurance and savings functions; first pillar to provide a minimum security against poverty; second pillar to provide savings via mandatory defined-contribution, fully funded schemes in the private sector; third pillar to provide additional voluntary savings for those who can afford it	Extend coverage; pluralistic pension system based in multiple functional tiers; the exact shape and form of tiers less important than the imperative that the system as a whole provide maximum security for older workers	Integrated and holistic strategy on ageing based on the life-course; basic income for all as the foundation for full socio-cultural inclusion and participation; active ageing strategies to avoid the costs of ageing; break down barriers to full labour market participation of older people (and all other disadvantaged groups)

Structure of policy conflict in the pension reform debate

How, then, does this triangular policy space structure conflict in the pension debate? The conceptual framework for policy-oriented discourse analysis introduced in Chapter 2 suggests that frame-based policy conflict of this type gives rise to asymmetric patterns of agreement and disagreement across advocacy coalition boundaries. Using the grid/group diagram to analyse these patterns shows that policy stories open up discursive spaces for potential agreement along either of the two dimensions. Thus probing the areas of agreement and disagreement means scrutinizing the potential for pairwise alliances across contending advocacy coalitions.

Areas of agreement: broad principles, common policy measures and mutual rejection

How do the affinities and mutually held disaffinities between the three contending discourse coalitions set up areas of agreement in the contested terrain of pension reform?

Advocates of both the crisis and stability stories agree that pension systems should promote rather than hinder economic growth and development. For this reason, economists advocating both the crisis story and the stability story agree on the necessity of a number of general pension reform measures. Since both argue that labour market distortions and labour supply disincentives hinder economic growth, economists from both the coalitions support policies to dampen any possible labour supply disincentives within the pension system; these typically include measures to tie benefits more closely to contributions (by retrenching redistributive elements in the benefit structure, by rigging the growth of benefits to demographic developments, etc.). Moreover, since spreading one's pension risks across a wide range of instruments makes good economic sense to proponents of either the crisis story and the stability story, neither the World Bank nor the ILO has any principled objection to private sector credit reserve, defined-contribution pension schemes. Further, since proponents of both the hierarchical and individualist policy stories perceive early retirement as a core policy problem, they can also (in principle) agree on statutory and socio-economic measures to raise the actual retirement age.

As a result, both the stability story and the crisis story are lukewarm about the egalitarian's preferred pension reform strategy: relocating the provision of income and services for older people into civil society. For the crisis story, this policy option is simply anachronistic: economic modernization is synonymous with the decline and dismantling of so-called 'informal' systems of provision (World Bank, 1994). Although organizations such as the ILO see a role for both civil society organizations and so-called 'micro-insurance' schemes in the provision of social welfare, these remain second- or third-choice policy options. Social welfare provision from civil society organizations – patchy and uncoordinated as it is – is nothing to shout about. Rather than signifying an increase in civic virtue, the growth of charitable social welfare provision reflects the scandalous dismantling of adequate statutory social protection (ILO, 2001, p95). Micro-insurance (e.g. schemes such as the Grameen Bank), in turn, are only suitable for workers in the developing world as a stop-gap since '... the level of resources generated is low and only limited social protection is provided' (ILO, 2001, p64).

Both the stability story and the social citizenship story emphasize the need for pension reform to strengthen 'social solidarity' and 'intergenerational solidarity' (European Commission, 1999; Gillion et al, 2000; ILO, 2001; European Commission, 2005).[5] For members of both types of advocacy coalition, social policy in general and pension reform in particular must aim at strengthening the elements of collective action and social responsibility. This is why advocates from both coalitions support measures that support people believed to be at risk of 'social exclusion'. In current pension debates, women and workers with discontinuous employment histories make up the marginal groups that constitute the focus for both egalitarian and hierarchical advocacy coalitions (these two categories overlap to a considerable degree). Hence, redistributive measures to boost pension claims for women who drop out of the labour market to provide care (either for children or older family members) are championed by both the proponents of the stability and social citizenship policy stories.

For this reason, advocates of both stories view pension reform proposals that threaten to embrittle these bands of social solidarity and social cohesion with the utmost suspicion. While the advocates of social citizenship concede that market-based approaches to social security have a place (albeit not a terribly prominent one) in an overall strategy of active ageing, the stability discourse warns that the implementation of individual savings accounts '... should not weaken solidarity systems which spread risks throughout the whole scheme membership' (ILO, 2001, p4).

Last, advocates of the crisis and social citizenship tales are likely to agree that pension systems need to promote choice and equality. For both coalitions, pension systems and social policy should be about creating the opportunities and the autonomy for individual self-determination. That is why both would argue that reforms need to rid existing old-age income provision of in-built inequities and unjust privileges. One way of doing this, proponents of either story argue, is to institute pension systems that provide universal flat-rate pension benefits independent of prior contributions (World Bank, 1994; European Commission, 1999, 2005; Bündnis 90/Die Grünen, 2002). Further, actors from both coalitions are strongly in favour of labour market flexibilization. Only flexible and efficient labour markets can provide the autonomy and choice for individuals to pursue their objectives.

Thus, advocates of the crisis and the social citizenship stories are both suspicious of highly stratified, complex and centralized pension systems. Calling for 'intergenerational fairness', both advocacy coalitions see these types of social security institutions as means of handing out favours to privileged minorities close to the centres of power – such as civil servants or male industrial workers or ruling cliques in the developing world (World Bank, 1994).

Intractable disagreement

Despite some overlap in principles and animosity, a three-way policy conflict is likely to be fundamentally intractable. Although members of contending advocacy coalitions may occasionally agree to disagree with the pension policy proposals of the third advocacy coalition in the debate, they disagree for different reasons. In the context of the international pension reform debate, this means that areas of broad agreement between the contending coalitions collapse when debate moves from general principles to more specific and practical policy implications.

Crisis vs stability

As we have seen, economists championing both the individualist and hierarchical stories readily agree that pension systems significantly affect economic growth. This, however, is where agreement ends. While advocates of the crisis story blame poor economic performance on the inherently distorting effects of pay-as-you-go, defined-benefit systems, economists in the hierarchical advocacy coalition deny that there is anything inherently wrong with the design of public pension provision (Orzag and Stiglitz, 1999; MacKellar, 2000). The only issue that is inherently problematic, they counter, is the inability of the crisis story advocates to produce any serious empirical evidence to back up their claims (MacKellar, 2000).[6] The alleged adverse economic impacts of public defined-benefit pay-as-you-go systems on labour markets and savings, they contend, are subject to considerable uncertainty. 'In most cases', the ILO argues,

> *... theory yields ambiguous predictions concerning these effects,*
> *empirical studies have failed to resolve the issues and controversy*
> *remains. However, there is little support for large effects of retirement*
> *benefit programmes in either labour and capital markets. (Gillion et*
> *al, 2000, p12)*

Further, while there may be general agreement that cuts to benefits are necessary, what needs cutting, by how much and when remains a hotly contested issue. Proponents of the stability story prefer cuts to maintain the institutional integrity of the pension system. Significantly, this involves maintaining public trust in the welfare state institutions; cuts, then, must be phased in gradually and managed judiciously by the experts. Advocates of the crisis story, such as the World Bank, point to the institutional integrity of current public pension systems as the cause of all the problems. The aim of any cuts and adjustments therefore is what the World Bank refers to as 'structural reform': a fundamental change of the nature of existing public pay-as-you-go, defined-benefit systems. This transition to multi-pillar pension systems, as the World Bank story demonstrates, needs to be rapid in order to allow as many people as possible to reap the economic benefits of reform.

Social citizenship and stability

The hierarchical and egalitarian pension policy stories express widely diverging definitions of the term 'social solidarity'. In general, the proponents of the stability story take the idea of social solidarity to mean the patterns of institutionalized and managed transactions between and within generations. These transactions have been pared down to simple and manageable flows of financial resources: here, solidarity is measured and expressed in monetary terms. Higher value transfers equate to more social solidarity. In contrast, advocates of the egalitarian social citizenship story perceive social solidarity in a far more essentialist and holistic manner. Although monetary transfers are one way of articulating social solidarity, the egalitarian story implies that true social solidarity requires real engagement in schools, in nursing homes, in individual households, in hospitals, at playgrounds and on the street (to mention but a few). Social solidarity is not only about the quantity of transactions that make up society but also about the quality of these exchanges.

For this reason, proponents of the stability and social citizenship stories are split on how best to build and foster social solidarity. In the language of the stability story, strengthening social solidarity means deepening and extending the implicit social commitments codified in the transfer structures of formal social security institutions. Thus, countering the threat of social exclusion for women and workers with discontinuous employment histories is a matter of extending coverage to unprotected workers and, if necessary, reconfiguring the transfer flows in existing social security machineries (Gillion et al, 2000; ILO, 2001). For members of egalitarian advocacy coalitions in the pension debate, in turn, dealing with demographic ageing requires more than simply expanding and fine-tuning impersonal social-security machineries. Rather, building and maintaining social solidarity, proponents of social citizenship maintain, means adopting a broad-based and integrated strategy in which income maintenance is one (albeit rather important) element (European Commission, 1999). However, tackling demographic change will require addressing and solving the deeper socio-cultural problems that relegate a large number of citizens (women, the old, the young, the disabled, ethnic minorities, etc.) to the margins of society. In short, we will need to do more than fine-tune large bureaucratic machines for transferring money: we will ...

> ... *have to invent new ways of liberating the potential of young people and older citizens. Dealing with these changes will require the contribution of all those involved: new forms of solidarity must be developed between the generations, based on mutual support and the transfer of skills and experience. (European Commission, 2005, p6)*

Crisis and social citizenship

Last, although both the individualist and egalitarian policy stories define large, centralized pension systems as a threat, the underlying perceptions of equality and choice in each account diverge sharply. For the proponents of the crisis story, equality refers to the opportunities and chances available to the individual. Here, the idea of enabling choice means providing an incentive structure that facilitates economic agents to make the 'right' choices, that is choices conforming to market rationality. Such an incentive structure is best provided in the private sector. The egalitarian notion of 'equality and choice', in turn, involves providing citizens with the autonomy to choose their preferred lifestyle at any point in the life-course. Significantly, this implies being able to freely move in and out of the labour market as life situations change without having to fear adverse effects on professional, social or family life. Institutions, particularly the market, ought to be geared to serve the real needs of citizens of all ages, not vice versa.

While individual policy actors from both advocacy coalitions favour flat-rate universal pension systems, their conceptions of the benefit level and role of these pension systems is very different. For advocates of the crisis story, a flat-rate universal system should be designed to alleviate poverty without distorting labour market supply decisions: that is, the pension benefits must be rather low to encourage labour market participation. Similarly, labour market flexibilization for proponents of the crisis story means breaking down barriers to the movement of labour; in practice, this amounts to the liberalization and dismantling of statutory labour protection to facilitate a more efficient clearing of the labour market. In contrast, proponents of the egalitarian social citizenship story see the flat-rate universal benefits as a basic income that severs the individual's dependence on the labour market. Thus, such a pension system (ideally an integral part of a basic income system for all citizens) would provide individuals with autonomy to choose to participate in the labour market or to pursue other socially valuable but consistently undervalued activities (such as child-rearing or care-giving). The flat-rate universal pension, then, is to provide all citizens with a basic level of material security that enables them to take an active part in socio-cultural life. On this view, labour market flexibilization aims to reshape the labour market to empower individuals to enter or leave the labour market when they wish. Rather than adapting the life-course to the requirements of work and career, egalitarian advocacy coalitions suggest that labour markets be made to adapt and cater for the human life-course. It is for this reason that the European Commission asks:

> *How can the organization of work be modernized, to take into account the specific needs of each age group? How can young couples' integration in working life be facilitated and how can we help them to find a balance between flexibility and security to bring up their children, to train and update their skills to meet the demands of the labour market? (European Commission, 2005)*

Table 4.2 summarizes the general areas of agreement and disagreement for the pension debate.

Table 4.2 *Structure of policy conflict about ageing*

	Crisis – Stability	Stability – Citizenship	Citizenship – Crisis
Agreement	Pension systems are important for economic growth	Concern for intergenerational solidarity	Universal, flat-rate pension system
Mutual Rejection	Extending ageing policy-making beyond social security institutions	Measures expose old-age income to financial risks	Status and in-built privileges in pension system
Disagreement	Precise role of pension systems in economic growth	Meaning of social solidarity	Level of flat-rate, universal pension

As in the transport policy debate, we can depict the policy space in which conflict about ageing and pension reform takes place in terms of a triangle (see Figure 4.1).

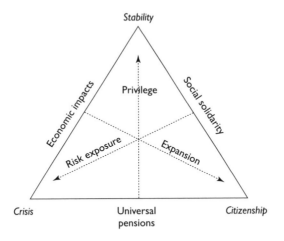

Figure 4.1 *Triangular policy space of the ageing debate*

Potential impacts of policy conflict in the pension debate

The three policy stories express policy frames that originate in the competing types of advocacy coalitions outlined in Chapter 2. Just as in the transport policy domain, individual policy actors in the pension reform debate use these policy frames to make sense of messy policy issues. By selectively highlighting specific aspects and backgrounding others, these frames act as interpretive templates for the construction of plausible and persuasive policy arguments. Since what is fore- and back-grounded is closely tied to a corresponding form of social organization, policy stories necessarily provide partial accounts of the demographic challenge and the pension reform issue. Each policy frame, then, has its peculiar strengths and weaknesses.

Crisis and risk
In the current pension reform debate, the policy-making initiative lies with the individualist advocates of the crisis story. The strengths of the crisis story are that it provides a powerful critique of conventional institutions of old-age income provision. Using demographic ageing as the linchpin, the advocates of the crisis story tell a compelling tale of inherent systemic flaws in public pension systems leading inevitably to inefficiency, waste, market distortions and financial collapse. The remarkable thing about this story is

that organizations such as the World Bank have successfully imbued the pension reform issue with a sense of urgency: although demographic and generational developments are among the slowest of social phenomena, the crisis story has successfully convinced policy-makers that time for reform is running out.

The crisis story's decisive weakness is the unquestioned assumption that markets by definition outperform public institutions in the provision of old-age income. This confidence in the market leads the advocates of the crisis story to underestimate the real risks and costs involved in multi-pillar pension systems. First, by shifting the long-term savings function into the private sector, the crisis story is exposing workers to considerable capital market risks (Wilmore, 1999). Second, receipt of a pension above the level of mere poverty alleviation requires continuous gainful employment in the labour market. For individuals unable to participate in labour markets (for reasons of illness and disability but also discrimination and marginalization), multi-pillar pension systems almost certainly lead to old-age poverty. Third, as we have seen, the crisis story systematically downplays the considerable transition costs incurred by the shift from existing public pay-as-you-go to multi-pillar pension systems. These costs could well offset any benefits expected from the structural pension reforms (Orzag and Stiglitz, 1999). Last, proponents of the crisis story consistently underestimate the importance of social security institutions for social cohesion and collective identity. In Europe, welfare states fulfil important socially integrative functions. On the one hand, the welfare state in general and pension systems in particular have prevented people at the top and the bottom of the income distribution from losing touch with society as a whole. On the other hand, European welfare states have become an integral part of national and regional identities in Europe (e.g. Nordic countries' identification with Nordic welfare states).

Stability and inertia

The hierarchical stability story tells a level-headed and more equivocal tale. In response to the challenge from the crisis story, the ILO and other organizations in the hierarchical advocacy coalition have tried to assure workers and citizens that everything is under control.[7] The strengths of the hierarchical pension policy story lie in the ability to bring a large body of expertise and technical competence to bear on the pension reform issue. This allows proponents of the stability story to question, qualify, find caveats, probe and, ultimately, deflate the more spectacular claims of the crisis story (see, for example, Orzag and Stiglitz, 1999). The upshot of their meticulous analysis is that the demographic and economic pressures on any type of pension systems are the same; successful old-age income provision, then, remains a question of meticulous design and ongoing management by competent experts.

This confidence in the expertise of a technocratic pension elite is the main source of weakness in the stability story's vision. Where the crisis story overestimates the ability of the market, the stability story places undue trust in the capabilities of pensions experts to rationally solve the challenges of demographic ageing. First, the preference for centralized and large pension administration based on complex statutory foundations makes for a rather unwieldy social policy instrument. This is even more inappropriate in times that require quick and effective policy responses. For example, in Europe, many pensions administrations have been sluggish to react to rapidly changing labour market structures (Ney, 2001). Second, entrusting pension management entirely to an expertocracy risks detaching pension policy-making from the social and economic realities (Nullmeier and Rüb, 1993). A related point here is that citizens no longer uncritically accept technical expertise as a legitimate reason for denying them participation in pension policy-making. Increasing levels of education have made citizens more discerning of experts and expertise (OECD, 2001). The expert-bias in the hierarchical stability story, despite many references to 'democratic management'[8] (Gillion et al, 2000; ILO, 2001), leaves little room for meaningful citizen involvement in social policy. Third, complex and highly stratified social protection systems – such as most continental European social insurance institutions – perpetuate social inequities and create barriers to entry into the labour market for marginalized groups.

Social citizenship and free-riding

Last, the egalitarian social citizenship story urges policy-makers to understand the challenge of demographic ageing in a holistic and pan-social context. Since demographic ageing will fundamentally transform all aspects of our societies, the policy debate about ageing needs to expand from its narrow focus on social security and health-care costs. The social citizenship story draws its strengths from understanding and addressing the wider socio-cultural contexts in which pension reform takes place. By thematizing the way demographic ageing will impact on key social systems, the members of the egalitarian advocacy coalition show how this fundamentally realigns socio-cultural and socio-economic relations. Unlike both the crisis and stability stories, the egalitarian social citizenship tale understands demographic ageing as a historically unique opportunity for restructuring society to suit real human needs (Walker, 2002).

The weaknesses of the egalitarian social citizenship story stem from an unwarranted belief in the robustness of community and citizenship. The communitarian bias implicit in the egalitarian policy story leaves the solutions open to a number of vulnerabilities. First, a comprehensive basic income system without bureaucratic sanctions or opportunity costs promotes free-riding. This, in turn, erodes the social capital and trust on which basic income systems are founded. Second, holistic yet decentralized social policy throws up some rather thorny public management issues. An integrated ageing strategy requires considerable coordination. Yet, as Rhodes argues, coordination and management are difficult in the types of polycentric policy networks envisaged by the egalitarian social citizenship story. Third, personalizing social policy by radically decentralizing provision for the communities also reintroduces an element of moral control and censure. Indeed, Alan Walker suggests that, in future, the rights to social services need to be balanced by obligations of citizenship: those who do not take care of themselves, either in terms of health or education, forfeit their entitlement to social services when old (Walker, 2002).[9]

Table 4.3 summarizes the frame-specific weaknesses and potential impacts of policy stories in the pension debate.

Table 4.3 *Potential impacts of policy conflict about ageing*

	Crisis	**Stability**	**Citizenship**
Trusts	Self-regulating powers of the financial market	Ability of social insurance institutions to solve demographic and economic challenges	Malleability of social norms and practices
Downplays	Financial risks to privatizing pensions	Inertia of social security institutions	Self-interest and greed
Vulnerable to	Creating and exacerbating old-age poverty	Stagnation and increasing social security costs	Free-riding

Conclusion

We have seen how three contending advocacy coalitions define the pension issue in different ways. By identifying and systematically comparing these policy stories about pension reform, the framework developed in Chapter 2 helped determine the *scope* of policy conflict in the contested terrain of pension reform.

In essence, all three policy stories tell us about the impacts of ageing on our societies and suggest ways in which we can best respond to these challenges. The individualist crisis story worries about the effects of ageing on economic growth. Shrinking labour forces and growing numbers of retirees will have adverse effects on the functioning of markets. This, proponents contend, is exacerbated by distortionary and inefficient pension designs.

The hierarchical advocates of the stability story, in turn, tell a different story. The fundamental problem is that most workers are not covered by old age social security and, for the few that are, these systems suffer from poor management. What is more, the structure of a particular pension system is not likely to make much difference to the impacts of ageing. Yet, scaremongering by vested interests in the financial and insurance sector have undermined trust in public pension systems. Worse still, they have managed to position credit-reserve, defined-contribution systems as the panacea for all of society's ills. The reality is, however, that both funded and unfunded, defined-benefit or defined-contribution face similar challenges: how to transfer more income from a relatively small working cohort to a large retired population. Solving this question still involves judicious management and is a job best left to pensions experts.

The egalitarian social citizenship story takes a more holistic and egalitarian approach. Pension reform, it maintains, is one element of an integrated policy strategy – called 'active ageing' – concerned with ageing populations. The real problems here are socio-institutional practices and attitudes that marginalize older workers and thereby accelerate the decline in the labour supply: reform in pension systems necessitates coordinated reforms in labour markets, health care systems, and social service provision. More importantly, however, Europeans need to break down and transform the discriminatory misperceptions that relegate older people to the margins of European societies.

All three policy stories tell us plausible tales about pension reform. They highlight certain aspects and background others. They set up a problem and provide the solutions. In doing so, they provide coherent and normative frameworks for thinking and arguing about pension reform. Together, these three stories delimit and delineate the contested terrain of pension policymaking: the scope of the three policy stories contains the ideas and arguments that make up the available pool of knowledge for policy-making.

This chapter has also explored the *structure* of conflict about pension reform. Despite inherent incommensurability, pairs of advocacy coalitions can agree on select general principles, general policy measures and common enemies. So, proponents of the crisis and stability story share the belief in the necessity of economic growth, the general efficacy of benefit retrenchments and the suspicion of community-based provision of social security. The advocates of the stability and social citizenship stories both understand the importance of social solidarity, support redistributive measures to strengthen social cohesion and abhor pension reforms that dissolve the interpersonal ties and obligations in favour or 'individual responsibility'. Last, the individualist and egalitarian types of advocacy coalitions agree on the desirability of choice and autonomy, on reforms that break down stratification and on the harm of complex, rule-based systems of social provision. However, these affinities and mutual rejections are not load-bearing structures: economists in individualist and hierarchical organizations cannot agree on how exactly pension systems affect economic growth, members of egalitarian and hierarchical advocacy coalitions have widely different notions of 'social solidarity', and the idea of 'choice' has rather different ramifications for individualist than for egalitarian groups.

What is more, the chapter has also shown that policy frames imply conceptual blindness. None of the policy stories can legitimately claim the possession of authoritative knowledge about the pension issue. Moreover, each type of advocacy coalition (either alone or in concert with another) identifies the inherent weaknesses of the others in the course of policy debate. Proponents of the crisis story do not easily perceive the real risks (capital market, disability, etc.) of shifting to a multi-pillar system. Advocates of the stability story are generally unwilling to see the sluggishness of publicly managed social security institutions as well as the secular decline in trust for expertise. Last, egalitarian social policy actors underestimate, simply discount or even welcome the impacts of free-riding on a system of social provision. Thus, like for the transport policy debate, a wide scope of policy conflict generates a rich reservoir or pool of ideas, concepts and strategies for dealing with the challenges of demographic ageing.

What, then, are the implications of this analysis for policy-making?

- First, the analysis has shown that policy debate in pension reform is fundamentally frame-based, contentious and intractable. A resolution by recourse to facts or by bargaining is not likely to be successful. Indeed, evidence suggests that incremental quid pro quo bargaining has not been overly successful in continental Europe (Ney, 2001; Bonoli and Palier, 2001; Leibfried and Obinger, 2001).
- Second, analysing structure of policy conflict shows that there are areas of agreement between the contending advocacy coalitions. However, agreement among all three advocacy coalitions is at a basic and general level (i.e. that demographic ageing is a policy challenge calling for policy action). Agreement on general principles and policy measures is more selective: here, the analysis shows an affinity between pairs of advocacy coalitions. Yet, agreement between advocacy coalitions is not particularly robust since contending policy actors define and understand the underlying issues in fundamentally different ways.

- Third, like in the transport policy debate, each advocacy coalition, if left unchecked, would implement pension reforms vulnerable to frame-specific unanticipated consequences. Thus, the analysis of this chapter also suggests that a lively policy debate with as many advocacy coalitions as possible may prevent consistent conceptual blindness. Again, policy-makers will need to find a way to resolve the dilemma of maintaining both a high degree of accessibility and a good quality of communication in the policy debate.

Notes

1   USA and Iceland at 2.1 as well as Mexico and Turkey at 2.2.
2   This indicator measures the share of the population aged less than 20 and more than 64 as a percentage of the 20–64 population.
3   The UN Population Division projects that, by 2050, the global median age, now at 28, will increase to 38 (UN Pop, 2007, p.xxvii).
4   Note that the savings function finds no mention here.
5   This is what we would expect since cultural theory's typology of advocacy coalitions suggests that the hierarchical and egalitarian positions on pension reform would prefer policies that foster and support collectives over the rights of individuals.
6   Even the World Bank acknowledges that the empirical evidence – particularly on the relationship between pension systems and the national savings rate – is inconclusive (World Bank, 1994).
7   Arguably with modest success. Pension reform has become a 'supercharged' policy issue (Walker, 2002).
8   Note that the ILO uses the term 'democratic management' not 'democratic participation'. Essentially, the ILO refers to bipartisan (i.e. employers' and workers' representatives) management of social security institutions. How and to what extent these forms of democratic participation reflect workers and, in turn, workers reflect the society as a whole, is an open question.
9   'Thus, the rights to social protection, lifelong education and training and so on may be accompanied by obligations to take advantage of education and training opportunities and to remain active in other ways' (Walker, 2002, p125). Although Walker immediately points out that this should not be used as a way to deny rights, it nonetheless implies that someone will be keeping score throughout the life-course.

References

Abel-Smith, B. (1993) 'Age, Work and Social Security: the Policy Context' in B. Atkinson and M. Rein (eds) *Age, Work and Social Security*, St Martin's Press, New York, NY

Bergheim, S., Neuhaus, M. and Schneider, S. (2003) 'Reformstau – Causes and Remedies', *Current Issues*, no 13

Bonoli, G. (2000), *The Politics of Pension Reform: Institutions and Policy Change in Western Europe*, Cambridge University Press, Cambridge, UK

Bonoli, G. and Palier, B. (2001) 'How Do Welfare States Change? Institutions and their Impact on the Politics of Welfare State Reform in Western Europe', *European Review*, vol 8, no 3, pp333–352

Börsch-Supan, A. (1999), 'Das deutsche Rentenversicherungssystem' in M. Miegel and A. Börsch-Supan (eds), *Gesetzliche Altersicherung – Reformerfahrungen im Ausland: Ein systematischer Vergleich aus sechs Ländern*, Deutsches Institut für Altersvorsorge, Cologne

Börsch-Supan, A. and Miegel, M. (1999) *Gesetzliche Altersicherung – Reformerfahrungen im Ausland: Ein systematischer Vergleich aus sechs Ländern*, Deutsches Institut für Altersvorsorge, Cologne

Brooks, S. and James, E. (1999) *The Political Economy of Structural Pension Reform*, papers.ssrn.com/sol3/papers.cfm?abstract_id=287393, accessed 5 February 2009

Bündnis 90/Die Grünen (1998) 'Solidarität neu begründen. Für eine gerechte und zukunftsfähige Gesellschaft: Bundestagswahlprogramm, www.gruene.de/archiv/wahl/btwahl98/prog/Wahlprog98/solidaritet.htm, accessed 21 March 2006

Bündnis 90/Die Grünen (2002) *The Future is Green: Party Program and Principles*, www.gruene.de/cms/files/dokbin/145/145643.party_program_and_principles.pdf, accessed 23 January, 2006

Esping-Andersen, G. (1990) *The Three Worlds of Welfare Capitalism*, Cambridge University Press, Cambridge, UK

European Commission (1999) *Towards a Europe for All Ages: Promoting Prosperity and Intergenerational Solidarity*, Com(1999) 221 Final, Brussels

European Commission (2005) *Confronting Demographic Change: A New Solidarity Between the Generations*, Com(2005) 94 Final, Brussels

Gillion, C., Turner, J., Bailey, C. and Latulippe, D. (2000), *Social Security Pensions: Development and Reform*, International Labour Office, Geneva

Gu, E. (2001) 'Dismantling the Chinese mini-welfare state? Marketization and the politics of institutional transformation, 1979–1999', *Communist and Post-Communist Studies*, vol 34, pp91–111

Hicks, P. (1997) 'The Impact of Ageing on Public Policy', *The OECD Bulletin*, December/January, no 203

Hill, M. (2007) *Pensions*, Policy Press, Bristol, UK

Hinrichs, K. (1998) *Reforming the Public Sector Pension Scheme in Germany: The End of the Traditional Consensus*, Paper presented at the XIVth World Congress of Sociology, International Sociological Association, Research Committee 19, Session 3: Reforming Public Pension Schemes (I), Montreal, Canada

Hinrichs, K. (1999) 'Rentenreformpolitik in OECD-Ländern. Die Bundesrepublik Deutschland im internationalen Vergleich', *Deutsche Rentenversicherung*, no 3-4/2000-06-06

ILO (2001) *Social Security: A New Consensus*, International Labour Office, Geneva

ILO (2007) 'Social Health Protection: an ILO Strategy Towards Universal Access to Health Care', *Issues in Social Protection: Discussion Paper*, no 19

Jiang, L. and O'Neill, B. (2006) *Impacts of Demographic Events on US Household Change*, IIASA Interim Report, no IR-06-030, Laxenburg, Austria

Leibfried S. and Obinger, H. (2001) 'Welfare State Futures: An Introduction' in S. Leibfried and H. Obinger (eds) *Welfare State Futures*, Cambridge University Press, Cambridge, UK, pp1–8

Liu, Y., Berman, P., Yip, W., Liang, H., Meng, Q., Qu, J. and Li, Z. (2006) 'Health care in China: The Role of Non-Government Providers', *Health Policy*, vol 77, no 2, pp212–220

MacKellar, L. (2000) 'The Dilemma of Population Ageing: A Review Essay', *Population and Development Review*, vol 26, no 2, pp365–397

Marshall, T. C. (1950), *Citizenship and Class and other Essays*, Cambridge University Press, Cambridge, UK

Metz, D (2002) 'The Politics of Population Ageing', *Political Quarterly*, vol 73, no 3, pp321–327

Ney, S. (2001) 'Country Report Germany', *Deliverable D2 – PEN-REF Project*, ICCR, Vienna

Nullmeier F. and Rüb, F. (1993) *Die Transformation der Sozialpolitik: Vom Sozialstaat zum Sicherungsstaat*, Campus Verlag, Frankfurt a.M.

OECD (1998) *Maintaining Prosperity in an Ageing Society*, OECD, Paris

OECD (2001) 'Government of the Future', *PUMA Policy Brief*, no 9, OECD, Paris

OECD (2006), *Society at a Glance: OECD Social Indicators*, OECD, Paris

Orzag P. R. and Stiglitz, J. E. (1999) 'Rethinking Pension Reform: Ten Myths About Social Security Systems', *World Bank Working Paper*, Washington DC

Pierson, P. (1994) *Dismantling the Welfare State: Reagan, Thatcher and the Politics of Retrenchment*, Cambridge University Press, Cambridge, UK

Pierson, P. (1996) 'The New Politics of the Welfare State, *World Politics*, vol 48, pp143–179

Pirages, D. (1997) 'Demographic Change and Ecological Security', *Environmental Change and Security Centre Report*, no 3.

Rehfeld, U. (2001) 'The German Pension Reform 2001', *PEN-REF Workshop –Setting European Pension Reform Agendas*, Laxenburg, Austria

Roush, W. (1996) 'Live Long and Prosper?', *Science*, vol 273, no 5271, p43

Schmähl, W. (2000) 'Pension Reforms in Germany: Major Topics, Decisions and Developments' in K. Müller, A. Ryll and H.-J. Wagener (eds) *Transformation of Social Security: Pensions in Central-Eastern Europe*, Physica-Verlag, Heidelberg

Seeleib-Kaiser, M. (2002), 'The Social Policies of the Red-Green Alliance in Germany', *International Journal of Political Economy*, vol 32, no 2, pp14–43

Sleebos, J. (2003), 'Low Fertility Rates in OECD Countries: Facts and Policy Responses', *OECD Labour Market and Social Policy Occasional Papers*, no 15, OECD Publishing, Paris

SoVD (2000) *Sozialstaat 2000: Überlegungen und Positionen zur Zukunft der Sozialpolitik in Deutschland*, Sozialverband Deutschland e.V, Berlin

Tullock, G. (1976) *The Vote Motive: an Essay in the Economics of Politics with Applications to the British Economy*, Institute of Economic Affairs, London

UN Pop (United Nations, Department of Economic and Social Affairs, Population Division) (2003) *World Fertility Report 2003*, United Nations, Department of Economic and Social Affairs, Population Division, New York, NY

UN Pop (United Nations, Department of Economic and Social Affairs, Population Division) (2007) *World Population Ageing 2007*, United Nations, Department of Economic and Social Affairs, Population Division, New York, NY

VDR (2000) *Rentenversicherung in Zeitreihen: Juli 2000,* Verband deutscher
Rentenversicherungsträger e.V., Frankfurt a.M

Walker, A. (2002), 'A Strategy for Active Ageing', *International Social Security Review,*
vol 55, no 1, pp121–139

Walker, A. and Naegele, G. (1999) *The Politics of Old Age in Europe,* Open University
Press, Buckingham

WHO (2002) *Active Ageing: A Policy Framework,* World Health Organization, Geneva

Wilmore, L. (1999) 'Public Versus Private Provision of Pensions', *DESA Discussion
Paper,* no 1, New York, NY

World Bank (1994) *Averting the Old Age Crisis,* World Bank Publications, Washington
DC

5

Health

Introduction

Health has gone global. Traditionally, health care provision, along with social policy or defence, has been the bread-and-butter of national policy-making. After all, people in the UK, Italy or Portugal rely on something called a national health system when they are ill. Increasingly, however, concerns about health – its state, its distributions or its management – have shifted to global policy arenas. Recently, the World Health Organization (WHO), an institution with a global perspective on health from the outset, has been joined in its efforts by other global policy actors, notably the World Bank (1993) and the World Trade Organization (WTO) (WTO Secretariat and WHO, 2002). Emergent health threats such as severe acute respiratory syndrome (SARS), pandemic influenza, bioterrorism and HIV/AIDS warn us that increasing economic interdependence and the global mobility that it engenders also create new and frightening vulnerabilities (WHO, 2006).

The relationship between health and globalization is anything but straightforward. One the one hand, peoples' health is still very much a local issue. Where we are born and whether we have to stay there in large part determines our likelihood of falling ill with, and of surviving, a particular disease. In this sense, the world is still a very large place with, it would seem, almost unbridgeable differences in health between different parts of the world. On the other hand, hyper-mobility seems to move not only goods, services and money, but also viruses, bacteria and other pathogens. Economic globalization, then, seems to have created highways that enable some diseases, notably SARS, pandemic influenza and HIV/AIDS, to cover vast spatial and social distance at breathtaking speeds (cf. Epstein, 2007 on HIV/AIDS in southern Africa). The global health crisis, it would seem, is what happens when disparate local worlds of disease and health get tangled in the continent-spanning nets of global hyper-mobility (see Chapter 3).

What, then, does globalization mean for health and health care provision? Much of the policy debate about globalization and health focuses on the so-called 'global health crisis'. This last case study, then, analyses policy conflict about the global health crisis. As in the previous chapters, the following section briefly outlines why the global health crisis gives rise to intractable policy controversy. The subsequent sections use the cultural theory framework to dissect this controversy in terms of its scope, structure and impact.

Global health crisis

The state of global health is complex and somewhat contradictory. In one part of the world – the rich, industrialized countries – advances in medical technology and institutional capacity have made an immense impact on societies. As we noted in the previous chapter, people now live much longer and healthier lives than their parents and grandparents. A girl born in Europe today can expect to live to be 77 years old, 68 of these years free of diseases and disability (WHO, 2007, p30). This is partly due to modern medical technology. The invention of antibiotics and other modern drugs has meant that communicable diseases, such as measles, cholera or tuberculosis no longer kill many people in developed countries: this is what is referred to as the 'epidemiological shift'. Moreover, the rapid pace of genetic research is helping doctors better understand and treat serious non-communicable diseases such as diabetes or cancer. Apart from medical technology, people in rich countries have also profited from effective institutions for delivering high-quality health services (OECD, 2004). Since the 1950s, most affluent countries, with the notable exception of the USA, have developed effective public health care systems. The advent of these institutions marked the beginning of a major public involvement in citizens' health (Blank and Burau, 2004); it is also at this time that public health became a defining activity of states.

In another part of the world – the poor developing countries – billions of people have no access to adequate health care. Poverty has meant communicable diseases are still the leading causes of mortality for people in the developing world. For example, in 2005 the WHO reported the prevalence of tuberculosis to be 511 per 100,000 of population in Africa, 290 per 100,000 in South East Asia and 206 per 100,000 in the Western Pacific. In Europe and the Americas, in contrast, tuberculosis prevalence measured at 50 per 100,000 and 60 per 100,000 respectively (WHO, 2007). Similarly, malaria, measles, pneumonia and diarrhoeal diseases account for just under 60 per cent of under-five infant mortality in Africa and 42.8 per cent of under-five infant mortality in South East Asia. In Europe, the same diseases only account for about 23 per cent of mortality of under-five-year-olds (WHO, 2007). Overall, the WHO reports that, in 2002, communicable diseases accounted for 83 per cent of all deaths in Africa and 55 per cent of mortality in South East Asia. Compare this to the Americas or Europe, where communicable diseases made up 27 per cent and 11 per cent of mortality respectively. All this means that people in the developing world live shorter lives with longer spells of disease and disability (WHO, 2007).

At face value, the explanation for these gross disparities in health outcomes is deceptively simple. People in rich countries are healthier and live longer because they spend considerably more on health care. For example, in 2003 Canada spent a thousand times more on health care per capita than Tanzania (at US$ 29 per capita), about 100 times more than China (at US$ 278 per capita) and about five times as much than South Africa (at US$ 669 per capita) (WHO, 2007).

The answer to the global health crisis, surely, is to increase spending on health in poor countries while maintaining the level of spending in rich countries?

But things turn out to be not quite as easy as all that.

Today, developed health systems face a far more complex set of health challenges than in the past. The success of modern medicine has fundamentally changed the nature of disease. Unlike many communicable diseases, it is difficult to pinpoint the causes of most chronic diseases. Science has identified a wide range of factors including sedentary lifestyles, diets rich in proteins and fat, and stress but also genetic predisposition and adverse living conditions that contribute to non-communicable diseases. Whether any particular individual falls ill to diabetes, heart disease or cancer, however, depends on the confluence of a very long list of behavioural, genetic and environmental factors.

More importantly, both the factors and their relative importance are hotly contested. Take, for example, obesity. As we saw in Chapter 3, some would argue that modern transport systems and contemporary land-use patterns are to blame for the sedentary life-style that causes obesity. The solution to obesity, then, is to fundamentally reform transport and land-use patterns to encourage more walking and cycling. Others argue that obesity is a genetic problem. Consequently, we should rely on technological innovation to come up with an effective therapy for obesity, possibly one that turns off the gene responsible for making us crave protein-rich and fatty foods (Meek, 2003). Others still argue that unhealthy food eaten with immoderation causes obesity; like cigarettes and alcohol, therefore, foods high in fat, protein and cholesterol need to be tightly regulated. Then there are those that question whether obesity and overweight contribute to the chronic disease burden at all (Oliver, 2006). While scientific research continues to generate new insights on risk factors, scientific knowledge has not been able to provide definitive answers in favour of any particular policy solution. The debate about obesity rages on.

Not only are the causes uncertain, but chronic diseases also require more complex and integrated approaches to therapy. The older people get, the more they tend to suffer from a range of diseases (a phenomenon called 'multi-morbidity') that require attention at different times (a phenomenon referred to as 'polypathology' or 'co-morbidity'). In an ageing population, therapy is less about cure than about disease management. This, in turn, calls for the integration of a wide range of competences in health and social services. At present, even the most advanced health systems are struggling to find effective solutions to these challenges (Walker, 2002).

Not least, dealing with chronic and non-communicable diseases is expensive. Diseases, such as diabetes or hypertension, incur higher costs for a longer period. Further, long-term care for the very old and frail is likely to increase health bills considerably as demographic ageing unfolds in the coming decades. Already, Americans spend 15.3 per cent of their GDP on health (OECD, 2007). In France and Germany, health spending accounted for 10.7 per cent and 10.3 per cent of national income respectively. In the member countries of the OECD, health spending is at an average of 9 per cent of GDP. Over the next three to four decades, the OECD projects that health care costs will grow by an average of 12.8 per cent if no cost-containing policy action is taken (OECD, 2006, p146). Many argue that these costs represent an intolerable burden on societies competing in an increasingly unforgiving global economy.

What about the developing world?

Emulating the evolution of health systems in affluent countries is unlikely to be a recipe for success here. Health systems in rich countries flourished in the sheltered confines of closed national economies. In contrast, health systems in the developing world today are exposed to the increasingly competitive global economy. Furthermore, as we saw in the previous chapter, demographic ageing is a global phenomenon. Middle-income countries are already feeling the health implications of uneven demographic ageing. In countries such as Malaysia, Thailand, China or even India, the growing prosperity of a burgeoning middle-class has generated what observers call the 'double disease burden'. In these countries, health systems are torn between the health needs of the poor, for whom communicable diseases remain the main cause of mortality, and the health demands of an affluent middle-class suffering increasingly from non-communicable diseases (WHO, 2000).

In addition, existing health care systems in the developing world suffer from a wide range of institutional ailments. In most middle-income countries, delivery systems are split into a well-endowed but expensive private sector catering to the affluent and an underfunded public sector health service that takes care of the poor. Typically, resources are inequitably distributed across the two competing sectors giving rise to considerable waste and inefficiency. In the poorest countries, delivery systems lack the most basic of supplies, such as beds or fuel for ambulances (Gilson, 1995). In most middle-income countries, fragmented financing mechanisms provide poor protection against the financial risks of disease (WHO, 2000). Financing of health care in the poorest countries is simply insufficient to cover even the most basic health needs of the population (WHO, 2000, 2006). Finally, poor governance practices – particularly corruption and incompetence – undermine any efforts of reforming these dysfunctional health systems. Box 5.1 shows how these flaws reinforce each other.

Box 5.1 *How multiple institutional flaws reinforce health system imbalances*

In many countries of the developing world, institutional dysfunctionalities tend to reinforce and reproduce each other. Most middle-income countries, such as the Philippines, Indonesia, Brazil and even India, health care financing is fragmented across a wide range of private and public sources. Typically, this means that the protection against the financial risks of disease are patchy and quality is poor. However, in principle, most middle-income countries have the institutional means for providing coherent financial protection. The problem is, however, that these institutions do not function as effectively as they could because of weaknesses in the governance practices: corruption and lack of basic regulatory competences are rampant in many of these countries. For example, the SUS in Brazil or PhilHealth in the Philippines – both social insurance type public contractors of health care – are struggling to bring private providers fully under their regulatory purview. Poor coverage, which undercuts the political and governance leverage of these institutions, makes this task all the more difficult. As a result, policy-makers cannot use the financing regime to steer delivery system development. Thus, delivery systems remain imbalanced, with public provision perceived (often accurately) as of lower quality than the more expensive private facilities. The wealthy avoid public health care systems in favour of private alternatives and since weak governance systems are incapable of compelling the rich to contribute, the public systems are starved of funds and remain unattractive.

Policy-makers in both the developing and the developed world have their work cut out for them. The latter will need to adapt existing high-performance health systems to new circumstances without compromising the coverage or quality of health care. Policy-makers in the developing world will need to expand their health systems under considerable competitive pressures. At the same time, they need to find institutional solutions for the changing nature of the disease burden. Health policy-makers across the globe, then, are embarking on a journey into the unknown for which there are no institutional templates or ready-made answers. This is why health policy debates give rise to intractable policy controversies.

Scope of policy conflict

Beyond the recognition that health care is getting ever more costly in the developed and increasingly inaccessible for most in the developing world, policy actors disagree about what this may mean for health governance. In coming to terms, quite literally, with these murky and opaque issues, actors tell three contending policy stories about health and health care provision. Each of these narratives sets out from different premises, goes on to identify different causes and, consequently, comes to very different conclusions about policy solutions.

Choice and health

In 2005, OECD countries spent an average of US$ 2759 per capita on health care, almost 9 per cent of GDP (OECD, 2007). In Singapore, in contrast, health costs made up 4.5 per cent of GDP (WHO, 2007). Remarkably, health outcomes and health status between these countries does not differ nearly as much as the difference in spending would suggest. The reason for this staggering difference, proponents contend, is not to be found in medicine or genetics but in simple economics. The predominantly public provision and administration of health care in both developing and developed countries has, unwisely, insulated health care provision from the harsh winds of competition and innovation. Based on an outmoded and unrealistic public-service ethic, public health systems have eroded individual responsibility in both health care consumption and provision. The outcome has been waste, inefficiency and inequality.

The proponents of the 'choices' policy story – which include international organizations such as the World Bank, libertarian think-tanks such as the Adam Smith Institute or the Economist, as well as market actors in the pharmaceutical industry – throw down the gauntlet to established health systems across the globe. Economic globalization, demographic ageing and technological progress, they argue, have made the bureaucratic and monolithic behemoths of public health care provision look tired and dated. What is needed, they contend, are health systems that can respond to the wants and needs of dynamic and competitive economies.

The setting: wealth means health

On this view, health and wealth are inextricably intertwined. Across countries, as Figure 5.1 shows, people in affluent countries are healthier than people in poor countries. Within countries, wealthy people tend to be healthier than the poor.

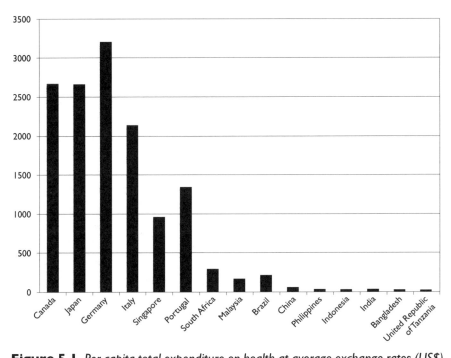

Figure 5.1 *Per capita total expenditure on health at average exchange rates (US$)*

This simple relationship, proponents argue, works as follows: on the one hand, health requires wealth. Economic growth generates the resources needed for developing the drugs that '... have brought huge benefits to the health and quality of life of millions of people over the last 100 years' (GSK, 2006). On the other hand, wealth requires health. Health industries drive economic growth by creating employment as well as buying goods and services (GSK, 2006; IFPMA, 2007). What is more, the products of pharmaceutical companies have '... directly and substantially contributed to economic and social development, by helping to build healthier and thus more productive societies.' (IFPMA, 2007, p7).

However distasteful it may be to some, so proponents of the choices story tell us, only firms pursuing profits in competitive markets can create the wherewithal to finance health care provision (IFPMA, 2007; GSK,

2006). So the hard-nosed reality for any health system – private, public or otherwise – is that economic growth ...

> ... *has a large impact on health outcomes – both by directly improving outcomes and by generating increased resources that can be mobilized by governments for increased public spending. (Gottret and Schieber, 2006, p4)*

According to this policy story, people make rational lifestyle choices according to their individual preferences: some prefer to smoke or engage in risky sports while others prefer to eat healthy food and keep fit. Since each individual is unique, there can never be a 'one-size-fits-all' conception of health and health care (Hansen, 2007). Good or bad health – how much of it we need at what time – is an intrinsic part of each individual, so much so that proponents of this story find the idea absurd that these health-related choices be left to someone other than each individual (Hansen, 2007).

The villains: waste, inefficiency and corruption

Health care provision in both rich and poor countries suffers from rampant inefficiency and waste brought about by politically motivated mismanagement (World Bank, 1993; Anonymous, 2004a, 2004b, 2004c; Hansen, 2007). While the symptoms may differ, so proponents of the choices story argue, the underlying disease of public sector ineffectiveness is the same.

In developing countries, the World Bank (1993) argues, the misallocation and inefficient use of scare resources creates gross health inequalities. Instead of funding low-cost but high-impact health interventions, policymakers waste resources on expensive but ineffective health services (World Bank, 1993). What people in the developing world need, so the argument goes, is affordable disease prevention such as vaccinations for common illnesses, safe food and water, as well as health education (World Bank, 1993). What they are getting instead are high-tech health services, more often than not funded by public subsidies, whose location and price are beyond reach for the people that need health care most (World Bank, 1993). The effects, as we have seen, are unnecessary suffering and excess mortality of the poor.

In the developed world, inefficiencies in-built into public health systems have sent health costs into orbit. The recent history of health sector reform in developed countries has been a protracted and largely unsuccessful battle against rising costs (OECD, 2004). This, proponents tell us, should not come as a surprise: reforms are little more than piecemeal adjustments that have not addressed the fundamental flaws of public health systems (Anonymous, 2004a, 2004b). As a result, health care expenditure has steadily grown over the past decades: since 1995 alone, health care expenditure has increased by an average of 4 per cent per year in OECD countries (OECD, 2007, p87).

The reason, proponents contend, that public health care systems are letting down patients around the globe is that they corrode individual responsibility (Anonymous, 2004d; Hansen, 2007). Socialized health care leaves those lucky enough to enjoy this privilege secure in the knowledge that, whether they smoke, participate in risky sports or allow themselves to become obese, someone else will foot the health care bill. Instead of taking responsibility for our health, Hansen (2007) argues, we rely on expensive therapies – paid for by others – to save us from our poor health-choices. Additionally, public health systems allow providers of health services to determine the 'need' and level of these services, a set-up that does not encourage restraint (Anonymous, 2004c). Providers, then, profit handsomely from irresponsible health behaviour. Thus, proponents argue, health systems reflect '... priorities of providers – doctors and hospitals – rather than consumers, and that there is plenty of scope to push for a better bargain from the huge amount of money now spent on medical care' (Anonymous, 2004c, p4).

In the developing world, pervasive corruption exacerbates these design flaws of government-dominated health care. Political elites use the public health system to divert much-needed resources from the poor to pander to rich but politically influential middle-classes (World Bank, 1993). As Daniel Vasella (2004), the president and CEO of Novartis, points out ...

> ... *poverty is aggravated by political mistakes, such as spending on military instead of investing in education, hygiene and preventive medicine. Some governments seem to care more about lining their own pockets than caring for citizens they are supposed to lead and represent. (p43)*

The heroes: empowering individuals, promoting choice

The solution, argue proponents of the choices story, is to administer a curative dose of individual responsibility by expanding patient choice. Creating what the World Bank calls 'health-enabling environments' calls for fundamental changes to existing health systems. It will require overcoming entrenched and powerful political interests. It will require overcoming the ingrained misperceptions that health care is best provided by the public sector. Most importantly, it will require convincing consumers and producers of health care that preventing diseases is considerably cheaper than curing them.

Health-enabling environments offer patients cost-effective health services. In the developing world, such environments target basic health services to those who need them most. This implies excluding those who can afford to pay for their own health care from public provision (World Bank, 1993). In order to free up much-needed resources, the World Bank (1993) argues, health policy-makers need to relocate health spending away from expensive HIV/AIDS treatment, curative treatment for terminal cancers or intensive care for very premature babies. These resources, argue proponents of this discourse, would be more wisely spent on a limited but cost-effective package of 'essential clinical services' for the poor (see Box 5.2). In the developed world, the standardization of common health services, so the argument goes, would go some way to create transparency about the costs of therapy. Ideas such as diagnosis related groups (DRGs) enable patients to clearly associate a set of health services with a particular price. Standardizing health services in this way, proponents contend, allows patients to choose the most suitable and cost-effective therapy.

Box 5.2 *Essential clinical services*

In their 1993 report, the World Bank lists the following minimum health care service that the public sector should make available to the poor.

- Pregnancy-related care
- Family planning services
- Tuberculosis control
- Control of STDs
- Common serious illnesses in young children
- Cover for accidents

Over and above service provision, health-enabling environments must encourage cost-effective choices on the part of the consumers. One way of doing this, proponents argue, is to provide information on health services and disease prevention. In the developing world, the World Bank (1993) points out, much can be achieved by strengthening basic education of the poor, particularly of girls. In rich countries, information technologies and the internet could enable patients to take responsibility for their health (Anonymous, 2004d; Hansen, 2007).

In addition to information, proponents of the choices story argue, health systems need to create compelling incentives for cost-effective health-choices. This implies a fundamental reform of existing financing mechanisms. On the revenue side, health care financing needs to reflect different risk profiles (Anonymous, 2004c; Hansen, 2007). Staggered insurance contributions or premiums financially reward people who choose to pursue a healthy lifestyle. In terms of expenditure, reforms will have to find ways to 'pay doctors for doing the right things at the right time' (Anonymous, 2004c, p18). Introducing more outcome-orientation in the payment of health care providers, proponents contend, will go some way to restoring individual responsibility. However, the most effective way of rewarding actual performance and quality, maintain many adherents of this policy story, is to abandon third-party purchasing of health care (Tanner, 2006; Hansen, 2007). In this way, consumers will have complete control and responsibility over their health care expenditure.

Taking responsibility for health choices presupposes that there are meaningful choices to take. In both the developing and developed world, advocates of the choices story see scope for far more competition and diversity in health care provision. Governments, so the argument goes, need to concentrate less on providing actual health care services and more on building strong, sustainable health markets. Box 5.3 outlines the contours of such a regime. The World Bank (1993) suggests that, in the developing world, privatization and decentralization in health service provision free up public resources for core tasks of poverty-alleviation and regulation. Decentralization of service delivery also expands choice: for this reason, policy-makers need to shift the emphasis of health care spending away from the expensive 'apex' (sophisticated tertiary care facilities located in cities) towards the base (basic primary care facilities located in the villages). Privatization and decentralization also generate choices, and encourage efficiency in the developed world. The Economist argues that,

> *Instead of trying to supplant the market, governments should be striving to promote competition while upholding social values about equity in health care. Thus, in Britain, a Labour government now sees no contradiction between its historic commitment to free services by doctors and hospitals and the pursuit of an internal market in the NHS. (Anonymous, 2004c, p18)*

Box 5.3 *Promoting medical innovation in health sector*

Effective regulation, proponents of the choices story contend, creates the right conditions for innovation and technological progress. Such a supportive environment, something the International Association of Pharmaceutical Manufacturers & Associations (IFPMA) calls the 'pharmaceutical innovation platform', consists of four basic elements (IFPMA, 2007; GSK, 2006). First, medical innovation needs strong and functioning health systems. Only strong health systems can create the right level of demand for pharmaceuticals to drive innovation. Second, medical innovation needs effective markets. This, so argues GlaxoSmithKline (GSK, 2006), implies '... greater liberalization in pharmaceutical pricing, especially for medicines that are not paid for by government' (p12). In any case, governments should avoid price controls and allow for market segmentation on a global scale. The profit-motive, unpleasant as it may seem to some, is the best incentive for developing effective drugs and therapies. Third, medical innovation needs strong IPR protection in order to ensure that profits accrue to those who have invested in costly medical research. Last, effective pharmaceutical R&D needs 'predictable and adequate regulatory requirements' (IFPMA, 2007, p25). Here, IFPMA (2007) suggests that ...

> ... *significant benefits could be achieved if medicine regulatory authorities applied a more pro-active and flexible approach to their processes and procedures, thus keeping up with the latest scientific and technological advances in the R&D process. (p36)*

In countries with no viable commercial markets for pharmaceuticals, so proponents of this story contend, there is very little that commercial firms can do on their own. Here, the onus is on the public sector to provide the right type of policy environment, making it attractive for pharmaceutical firms to invest. Policy actors, the IFPMA contends, should be looking to design and implement '... mechanisms that can provide incentives that can compensate for the absence of a viable commercial market for medicines' (p52). One way of doing this is to use cooperative projects, so-called joint public–private initiatives (JPPIs) or public–private partnerships (PPPs). Here, firms, states and NGOs collaborate in making medicines for specific diseases more accessible to the poor in the developing world.

People's health depends on their wealth. That is why, the World Bank reminds us, '... [e]conomic policies conducive to sustained growth are thus among the most important measures governments can take to improve their citizen's health' (p7). Health policy-makers around the world would do well in understanding the implications of this simple truth. 'What is needed,' The Economist tells us,

> ... *is a change in the structure of health care systems so that competitive pressures push them in a more useful direction, enhancing power of purchasers and increasing competition in the supply of medical care. The traditional argument has been that health care is too important to leave to the market. The opposite holds true: it is too important not to be exposed to the market. (Anonymous, 2004c, p17)*

Health is a human right: an egalitarian story

Consider the following! People in the developing world today suffer and die unnecessarily because they are poor. Poverty prevents them from acquiring basic medical treatment for diseases such as dysentery or malaria. At the same time, people in the developed world suffer and die unnecessarily because they are rich. Profligate lifestyles that are unhealthy and unsustainable are slowly killing the increasingly sedentary and obese population in the developed world. Something, surely, has gone horribly wrong.

For the proponents of the 'health rights' story – NGOs, humanitarian organizations, new social movements but also critical physicians – only the deluded or malicious fail to see how perverse the state global health is today. The health care mess we are in, they argue, is beyond technocratic fixes or market utopias. What we need – and we need it quickly – is fundamental change in the way we live.

The setting: health is more than the absence of disease
The keystone of the health rights story is the Alma-Ata Declaration of 1978. In outlining a comprehensive primary health care reform agenda, the Declaration defines health as ...

> ... *a state of complete physical, mental and social well-being, and not merely the absence of disease or infirmity, is a fundamental human right and that the attainment of the highest possible level of health is a most important worldwide social goal whose realization requires the action of many other social and economic sectors in addition to the health sector. (p2)*

This broad definition lays out the basic beliefs of the health rights story. First, health is about a lot more than just disease. Understanding health, so the argument goes, requires looking at the full range of factors that shape human well-being. Apart from the prosaically biological, then, '... *the main determinants of health are social, economic and political*' (Werner, 2003, p17, original emphasis).

Second, since all humans are equal, so too are their needs for health and well-being. This is why the PHM (2000) pronounces that the ...

> ... *attainment of the highest possible level of health and well-being is a fundamental human right regardless of a person's colour, ethnic background, religion, gender, age, abilities, sexual orientation or class. (p3)*

It follows, proponents contend, that health is neither negotiable nor can it be traded. Health and well-being cannot be parcelled off and sold just as access to them cannot be rationed and prioritized. So fundamental is this right to health and well-being that proponents of this story believe it ought to be the primary focus for policy-making (PHM, 2004).

Third, this type of health implies holistic and participatory health policy-making. Health policy that does not address social, economic and political conditions can only ever alleviate with the symptoms (Sanders, 2003). Effective health policy will need to integrate processes in a range of different policy domains, such as the environment, transport or industrial policy. Significantly, comprehensive and needs-oriented health care must integrate a broad spectrum of actors, including patients, traditional healers and critical social groups (WHO and UNICEF, 1978; Sanders, 2003; Werner, 2003).

The villains: inequality and dependence

Of the billions of US dollars spent on medical R&D each year, so Oxfam, VSO and Save the Children (2002) inform us, only a mere 10 per cent is committed to diseases that account for 90 per cent of the global disease burden (p19). It is these screaming inequities, the outcomes of an obscenely unjust and exploitative world order, that cause disease and suffering around the world.

Economic globalization has commodified and demeaned health (Banerji, 2003; Werner, 2003; Sanders, 2003). As a result, so the health rights story argues, billions of people on the planet today have no access to adequate health care. To make room for commercial health care provision, so proponents of the health rights story argue, neo-liberal economic policies, such as structural adaptation programmes in the south, and massive cuts to social budgets in the north, have all but destroyed public health systems (Banerji, 2003; Werner, 2003; Sanders, 2003). The perpetrators of this crime – political elites beholden to powerful business interests –

have tried to conceal the dismantling of public health behind unfulfilled promises of efficiency, choice and quality (Schiff et al, 1994; PHM, 2000). Consequently, the ability to pay for health care has crowded out the social right to health.

Exacerbating global and local inequities, argue proponents of the health rights story, invariably punishes the weak and vulnerable. In countries such as India or Bangladesh, de facto privatization of health services has closed down access to the most basic of health services for the vast majority of people (Islam and Tahir, 1999). Additionally, proponents contend, local health inequities grow as health becomes a factor in macro-economic and corporate production functions. The more health care focuses on the economically active, the more health needs of children, the old, the disabled or pregnant women fall by the wayside (PHM, 2000; Benson, 2001; John, 2003).

The profit motive also corrupts health and the quality of care for people in rich countries. Even if you are not among the unlucky 39 million US citizens without health insurance cover, all the exorbitant health insurance premiums buy you is dependency on high-tech drugs, flashy diagnostic gadgetry and macho surgical procedures (Robinson, 2008, p1). The dominant curative approach to health care does little for your real health needs, but works wonders for 'shareholder profit, inflated CEO salaries, and top-heavy administration' (Bündnis 90/Die Grünen, 2002; Robinson, 2008, p1). The American organization Physicians for a National Health Program (PNHP) argues that quality …

> … *is distorted when the ability and willingness to pay become the criteria for determining which services are provided. Marginally effective or even harmful treatments for the well-insured affluent take priority over more needed and appropriate services. (Schiff et al, 1994, p3)*

If poverty that is caused by economic inequity is the '… mother of all health problems' (Sivaraman, 2003, p21), then the father is political exclusion. Globally, so the health rights story argues, a 'syndicate of rich countries and the ruling elite of poor countries' (Banerji, 2003, p12) have scuppered attempts at implementing Alma-Ata's revolutionary and transformative vision of health care. Well organized and well funded, these forces have ruthlessly used their dominant positions in government, business and research to smother equalitarian ideas of health care. In their place, this cartel has installed the '… dominant inequitable paradigm of development' (Werner, 2003, p18). Box 5.4 shows how the 'cartel' has undermined equitable and humane health care provision.

Box 5.4 *The cartel and PHC*

One way the 'cartel' flexed its immensely powerful muscle, proponents of the health rights story argue, was by reinterpreting the PHC agenda. The vision of primary health care was usurped first by the Johns-Hopkins School of Public Health. Researchers of this august institution, so the argument goes, degraded the inclusive approach of primary health care to selective interventions that empower technocratic elites (Werner, 2003). It was then further eroded by the structural adjustment programmes that razed to the ground the fledgling public health institutions to make room for commercial providers. Soon thereafter, so the story goes, the World Bank appropriated the health policy issue. Their 1993 report urges health policy-makers in the developing countries to further slash public health budgets and allow more private provision, all in the interests of greater equality, of course. The last and probably most depressing instance in this horrific story, proponents lament, is the recent sell-out of the WHO and UNICEF. Under immense pressure from the 'cartel' and 'syndicate' elites, these organizations have endorsed so-called public–private partnerships with ethically dubious firms and corporations, such as junk food chains and weapons manufacturers.

Locally, the dominant interests of providers marginalize important stakeholders, such as patients, family members or non-medical health professionals. Exclusion from policy-making more often than not leads to discrimination of marginalized groups, such as the very old, but also people of a different sexual orientation, in commercial and technocratic health systems (PHM, 2000; Bündnis 90/Die Grünen, 2002). As a result, the International Alliance of Patients' Organisations (IAPO) contends that ...

> ... *the patient's voice is not valued enough in policy-making and practice. Patient involvement is often merely tokenism; its influence on policy-making can be restricted by practical and financial structures, differing knowledge bases, cultural barriers and personal attitudes. (IAPO, 2005, p1)*

The global economic and political system, argue proponents of the health rights story, is as 'undemocratic as it is unsustainable, [since] it promotes economic growth of the rich regardless of the human and environmental cost' (Werner, 2003, p17). Instead of liberating and empowering people, commodified health care imposes inappropriate 'cures' on both the poor and the rich.

The heroes: empowerment and liberation

Unfortunately, proponents argue, there are no quick fixes, no miracle cures and no technical gadgets to get us out of this mess. A system that lets millions of poor children die for lack of cheap medicine while marketing expensive breast enhancements to rich teenagers is way beyond reform. A system in which the poorest have to pay the highest prices for health care is rotten to the core.

Fighting the global disease burden means tearing down the social, economic and political barriers to equitable and empowering health care. In both the developing and developed world, proponents of health rights demand free and universal access to primary health care (PHM, 2000; Bündnis 90/Die Grünen, 2002). Since the private sector cannot be trusted with this task, proponents contend, this is a central responsibility of the public and tertiary sectors (PHM, 2004). Further, equitable and empowering health care, so the argument goes, is best provided at local level (Sanders, 2003). In both the developing and the developed world, advocates of health rights call for the radical decentralization of health care provision (PHM, 2000; Bündnis 90/Die Grünen, 2002). Moreover, inclusive and empowering health care, proponents argue, must appreciate the holistic relationships between people, their environment and their health. This means 'demystifying health and medical technologies' (PHM, 2000) by integrating traditional medicine into mainstream allopathic therapies. Most importantly, empowering health care requires a 'strongly participatory strategy' (Werner, 2003, p15). For this reason, civil society and communities, proponents contend, must play a central part in health policy processes (Sanders, 2003; Werner, 2003; Bündnis 90/Die Grünen, 2004).

However, argue the proponents of this discourse, changing health care is only part of the story. If reforms are to treat more than just the symptoms, then they must address global economic and political inequities. First, the reckless casino capitalism and its institutions – the WTO, the World Bank, the IMF and the Security Council – must go (PHM, 2000). In particular, the WHO needs to radically change to be able to respond adequately to the health needs of the poorest by insulating them from the commercialization of health care (PHM, 2000). Second, in order to end the ruthless exploitation of the weak and poor, global economic processes must be brought under democratic control. This not only implies the regulation of multi-national corporations but also the taxation and control of global financial flows. Third, the growth-oriented paradigm, as dominant as it is wrong-headed, must give way to a renewed commitment to 'equitable social investment' (PHM, 2000). Last, an equitable world order must put an end to the wholesale destruction of nature caused by profligate consumption of the rich (PHM, 2000; Bündnis 90/Die Grünen, 2002). That is why the People's Health Movement calls for activists to pressure 'wealthy and industrialized countries to reduce their consumption and pollution by 90 per cent' (PHM, 2000, p6).

A just world order must rest on social, political and economic equality. All other foundations, the proponents of the health rights story argue, inevitably lead to exploitation, suffering and disease. For this reason, the People's Health Charter proclaims:

> *Equity, ecologically sustainable development and peace are at the heart of our vision of a better world – a world in which a healthy life for all is a reality; a world that respects, appreciates and celebrates all life and diversity; a world that enables the flowering of people's talents and abilities to enrich each other; a world in which people's voices guide the decisions that shape our lives. (PHM, 2000)*

Stewardship and health: a hierarchical story

Providing health care is a complicated undertaking. It involves deploying a veritable army of physicians, nurses and pharmacists. It means getting the right resources to the right places at the right times. It calls for anticipating future needs and trends. More importantly, health professionals must ensure that patients are guided through the health system with care. Since people are likely to return to the health system at some point, effective provision calls for meticulous record-keeping. Whatever else it may be, providing effective health care is always an immense organizational and administrative challenge.

In the past, so the proponents of this story contend, only the type of administrative skills found in public organizations have been equal to this challenge. And, they continue, this will remain the case in the foreseeable future. Neither the visions of glitzy high-tech medicine nor the touchy-feely PHC utopias can dispel the fundamental truth: health care needs competent administration, oversight and, yes, not a little control.

The setting: striking balances

Health, advocates of stewardship contend, is an unusual thing. In certain respects, health is a commodity. Not only, the European Commission (2007) tells us, is health '... important for the well-being of individuals and society, but a healthy population is also a prerequisite for economic productivity and prosperity' (p5). And yet, unlike a pair socks or a car, health affects us in an uncomfortably immediate and inescapable way. After all, the consumer of health care '... is also the physical object to which all such care is directed' (WHO, 2000, p4). Moreover, unlike assets such as education, health is subject to large and unpredictable risks that are independent from one another. Since many health choices are made in the context of fear, pain or bereavement, people cannot be expected to make decisions about health like they would about, say, living-room curtains (Blunden and Smith, 2005; BMA, 2004). This is, proponents argue,

> *... why markets work less well for health than for other things, why there is need for a more active, and also more complicated, role for the state, and in general why good performance cannot be taken for granted. (WHO, 2000, p4)*

In other respects, then, health also is a right. Apart from fulfilling the health care wants of those who can afford it, health systems 'have an additional responsibility to ensure that people are treated with respect, in accordance with human rights' (p4). And yet, like any other right, health is not an absolute right but competes with other human rights, such as freedom of speech or freedom of mobility (Hunt, 2004). In each instance, health rights need to be balanced with other competing rights. For example, health policy-makers need to encourage people to adopt healthy practices without infringing on the right of individuals to self-determination.

For the advocates of the stewardship story, then, health care is about striking a balance between health as a commodity and health as a right. Box 5.5 shows how the WHO has operationalized this task. Hierarchical policy actors know that a balance between these potentially contradictory forces does not occur spontaneously. Markets, the advocates of stewardship argue, are very good at responding to consumer-driven wants of the population. However, proponents argue, markets are incapable of addressing the issue of fairness and equity; as a result, health care is good only for those who can afford it. Similarly, district health systems and primary care are geared to fairness and quality. Primary health care approaches tend to subsume wants and expectations of the population, however, to the (perceived) real health needs of the poor, vulnerable and marginalized. A balance between these forces does not fall from the sky – it needs to be brought about.

Box 5.5 *WHO health system goals*

In the World Health Report 2000, the WHO lays out the fundamental objectives of effective health systems. Successful health care provision, the WHO (2000) argues, involves balancing and optimizing three goals.

First, health systems, obviously, should improve health. Not only should health systems cure disease and ameliorate accidents when they occur, but they should also actively prevent disease and accidents from occurring.

Second, health systems also need to respond to the expectations of the population. Since health services are a commodity, health systems must be responsive to the kinds of therapies people demand.

Third, effectiveness in health systems ensures the 'fairness of financial contribution'. Since health care provision is also a right, health systems must ensure fair access to health services for all that need them.

Concentrating solely on one or two of these goals, proponents argue, will lead to an imbalance of the health systems. A health system geared only towards satisfying health demands will concentrate scarce health resources on those willing to pay the highest price. As a result, this will close down the access to health care for those whom this price is too high. Similarly, a system focused on fairness alone will alienate those who are willing to pay for distinction in health care provision. Finally, a system that concentrates solely on the technical aspects of health service provision will become unresponsive and unwieldy.

The art of striking these balances is called Stewardship. Saltman and Ferrousier-Davis (2000) define stewardship as a '... function of government responsible for the welfare of the population, and concerned about the trust and legitimacy with which its activities are viewed by the citizenry' (p735). It involves, the WHO (2000) tells us, '... the tasks of defining the vision and direction of health policy, exerting influence through regulation and advocacy, and collecting and using information' (p.xiv).

The villains: imbalanced health systems

On this view, health policy challenges in the developing world are diametrically opposed to problems in industrialized countries. In the developing world, health systems are out of balance. As a result, they are not keeping people healthy, cannot meet their demands or provide fair finance. In countries of the developed world, by contrast, ageing and globalization ask awkward questions of health policy-makers that, if not answered with the utmost care, threaten the balance of these high-performance health systems.

The root cause of health system imbalance in the developing world, proponents contend, is the institutional disease of poor stewardship. First, the WHO (2000) diagnoses, poor stewardship impairs health policy vision. Policy-makers lose sight of the population's needs and demands. Poor stewardship means that policy-makers are unable to 'see' and engage with other important actors, such as the private sector. Often, poor stewardship can lead to an inability or unwillingness to see corruption and malfeasance; as a result, '... stewardship is subverted; trusteeship is abandoned and institutional corruption sets in' (WHO, 2000, p121).

Second, poor stewardship can lead to policy paralysis. The WHO points out that many developing countries have no explicit policies for the health sector. In countries that have formal health policies, these are often inappropriate for the given level of capabilities and resources. Policy paralysis also closes down important pathways of health system development. For example, while privatization can create more capacity for health care provision, it presupposes significant regulatory capacities. In countries with weak stewardship, privatization cannot make this positive contribution to provision; instead, it exacerbates imbalances and inequities (WHO, 2000).

Third, blinded and immobile, the patient becomes vulnerable to political abuse and exploitation. Weak stewardship leaves health systems exposed to the irrationality typical of politics in poor countries. Box 5.6 describes what poor stewardship can do to health systems.

In the developed world, in turn, so proponents of the stewardship story point out, health systems are, on the whole, well-balanced. And yet, demographic ageing threatens the balance of developed health systems. As we have seen, ageing populations demand more and more expensive health services (OECD, 1998; European Commission, 2007). Inopportunely, this is happening at a time when globalization is squeezing public coffers. Maintaining performance will require health policy-makers to rebalance, recalculate and redistribute the societal burdens of health care provision.

Box 5.6 *Features of imbalanced health systems*

The institutional disease of poor stewardship, the WHO (2000) argues, presents itself as follows. Afflicted health systems show one or more of these symptoms:

- health systems are highly bureaucratic;
- health systems are poorly managed;
- the structure of health care provision is centralized and hierarchical;
- health systems are fragmented by vertical programmes in which areas are operated like fiefdoms;
- health systems are overly dependent on uncertain donor funding (WHO, 2000, p120).

Regrettably – and this, proponents contend, is what really afflicts developed health systems – governments are coming to the wrong conclusions. In the past two or three decades, health reforms in affluent countries have tried to control health care costs by introducing competition (Freeman, 1998). This narrow focus on competition, however, undermines the basis for effective health care planning and management (BAK, 2003; BMA, 2004). The BMA (2004) contends:

> *In current policy, patient choice is driven by a belief that making hospitals compete with one another in a market for patients will result in providers shaping services the way that patients want them so they will come to their institution. The BMA is very concerned with some assumptions in this view. We do not believe that the difficult financial environment that will ensue will be conducive to providers taking a long term strategic focus or concentrating on organizational development. (p1)*

What is more, competition in health systems actually reduces choices by destroying capacity (BMA, 2004; BAK, 2008). On the one hand, health care providers who cannot 'attract' customers face closure. In this way, competition '... will destabilize a vital public service and lead to inequalities in service provision from area to area that will fragment care and be very costly' (BMA, 2004, p6). In Germany, reforms have led to hospital closures, the rationing of health services, and the geographical concentration of facilities (BAK, 2004). On the other hand, 'ruinous competition' (BAK, 2003), has undermined the autonomy of health professionals by empowering public and private health care insurers. Cost controls by insurers spell the end of 'individual needs-oriented patient care' (BAK, 2004, p8). What was a confidential therapeutic relationship is now a three-way transaction, with insurers sitting at the apex. Accountancy, the legal profession and, worse still, politics eclipse medical skill in determining permissible diseases and appropriate therapies. Invariably, the quality of health care suffers since targets and standards '... that are unrelated, too numerous and politically driven distort clinical care' (BMA, 2004, p12). In the end, physicians are downgraded from autonomous professionals to mere providers of predefined and standardized medical services. As a consequence, argue proponents of the stewardship story, health professionals are turning away from health systems in which they are forced to pick up the tab for poor management and persistent underfunding (BAK, 2008). This, in turn, exacerbates the acute shortage of qualified health professionals in the developed world. How this contributes to more patient choice, proponents of stewardship ask, remains a mystery.

The heroes: restoring balance

Health policy-makers around the globe have their work cut out for them. In the developed world, health systems must start providing basic health services to all who need them while finding ways of financing health care that protect the vulnerable (WHO, 2000; ILO, 2007). This calls for the expansion of existing capacities as well as the more efficient use of existing resources. In health systems of the developed world, the issue is how best to maintain the high quality of health care in the face of changing demographic and socio-economic circumstances.

Before health policy-makers can turn to the market or to the community, they need to get stewardship right. Stewardship enables policy-makers to balance the many things in health care provision that pull in opposite directions: things such as public and private interests, compassion and hard-nosed technical competence, choice and guidance, or patient rights and professional autonomy.

And because stewardship must square these contradictory pressures, '... the ultimate responsibility for the overall performance of a country's health system must always lie with government' (WHO, 2000, p119). Governments are the only societal actors, so the argument goes, at the right vantage point to ensure 'coherence and consistency' across different departments and sectors (WHO, 2000). This, proponents of the stewardship story argue, does not necessarily mean a return to bureaucratic planning of health care. On the contrary, stewardship recognizes that health policy-makers operate in increasingly fluid and dynamic environments. Effective stewardship, the WHO (2000) argues, implies that states' '... key role is one of oversight and trusteeship – to follow the advice of "row less and steer more"'(p119).

Rowing less and steering more implies three practical tasks. First, good stewardship must build robust institutional vessels for health care provision. Stewardship must clear a legal, political and socio-economic space in which other policy actors can deliver health care services (ILO, 2007). In the developing world, creating these spaces means fighting persistent corruption and malfeasance to insulate health policy-making from the irrationalities of politics (WHO, 2000, p.xvi). Responsible and constructive policy deliberation, so the argument goes, is what the International Labour Organization (ILO) likes to call a 'social dialogue': deliberation between accredited stakeholders in a 'prepolitical space' (ILO, 2007). Good stewardship also needs strong administrative institutions that can take an active role in shaping health care provision as well as health needs (see Box 5.7).

Box 5.7 *Stewardship and disease prevention*

Of course, argue advocates of Stewardship, the best way to reduce health care costs in an ageing society is to prevent these costs from even occurring. The European Commission (2007) estimates that ageing is likely to ...

> ... *push up healthcare spending by 1 to 2% of GDP in Member States by 2050. On average this would amount to about a 25% increase in healthcare spending as a share of GDP. However, Commission projections show that if people can remain healthy as they live longer, the rise in healthcare spending due to ageing would be halved. (p7)*

Good stewardship, then, also involves actively preventing disease and promoting health. Effective prevention, proponents argue, is based on the dissemination of reliable information and knowledge to the relevant stakeholders (WHO, 2000; BAK, 2003). However, given the profound social changes that ageing is likely to bring about, the institutions that finance health care provision will have to adjust accordingly. The International Social Security Association (ISSA) foresees that social insurance institutions will need to change ...

> ... *from a more reactive role as payer/purchaser of curative, rehabilitative and long term care services to a more pro-active role to motivate and engage the insured population and the actors in the social health insurance system framework in health promotion and prevention. (ISSA, 2007, p3)*

Far from playing a smaller role in health care provision, the ISSA see the future of real prevention as involving social insurance institutions becoming a proactive partner in the health promotion of each individual. Much like the physician as a navigator, this vision sees experts and health professionals in health insurance institutions as facilitators, advisors and guides to information and services about health promotion (ISSA, 2007).

Second, good stewards need to chart a course for the vessel. Health policy-makers, the WHO (2000) asserts, have the responsibility of defining the vision and determining the general direction of the health system. For the German Medical Association (BAK), this means that governments '... determine the type, the volume and the level of social protection in case of disease and in terms of prevention' (p2). Instead of formulating elaborate and detailed health sector plans that cannot be implemented, health

Table 5.1 *Scope of policy conflict about the global health crisis*

	Choice	*Rights*	*Stewardship*
Setting	Wealth means health; health is what you think it is; health care provision is about satisfying health wants	Health is a right; broad definition of health; health care provision is about meeting basic health needs	Health is not quite a commodity and it is not quite a right; health care provision is about striking a balance between contradictory tendencies
Villains	Public health care systems do not encourage efficient consumption of health goods; not only ineffective but also unfair; poor countries: inefficiency, misallocation of resources, corruption; rich countries: skyrocketing	Global inequality causes disease; inequitable economic system has eroded access to health for billions; inequitable political system; exclusion at all levels	Different worlds, different problems; poor countries: health systems are imbalanced because of poor stewardship; rich countries: balanced systems are being undermined by ideological commitment to choice
Heroes	Create real choices in health care provision; health enabling environments; empower consumers viz. providers (information, education); right incentives for individual responsibility; encourage diversity and plurality	Eliminate inequities; break down financial, geographical, ideological and political barriers in health systems; new global governance; new global development paradigm	Strengthen stewardship; steer more, row less; create appropriate public sector institutions: insulate from politics; processes for goal setting; skilful day-to-day management: education, management of networks, guidance; good regulation
Moral	Get the prices right	Primary health care for all	New universalism

policy-makers in developing countries should concentrate on designing flexible health policy frameworks (WHO, 2000).

Third, in addition to navigation, good stewardship requires skill at the helm to hold the course from day to day. This, argue proponents, means exerting the right kind of influence on providers and patients through coordination, regulation and education. This involves, the WHO (2000) warns, striking '... a balance between avoiding regulatory capture by private interests and maintaining productive dialogue with them to ensure that regulatory frameworks are realistic' (p129). For the European Commission (2007), good stewardship means identifying and managing the synergies between health systems and other policy sectors (p6).

Similarly, the future role of health insurance, the ILO (2007) argues, involves the management of networks and intersectoral partnerships: here, '[c]oordination and collaboration are key in a multi-actor environment and crucial for avoiding duplication and inefficiencies' (p6).

Effectively tackling the current global health problems, advocates of the stewardship story conclude, will require a new mindset. Leaving health care entirely to the market or to holistic participative primary care, the stewardship story tells us, is a recipe for failure. What is needed today in both the developing and the developed worlds is an approach that exploits the synergies of both extremes while avoiding their pitfalls. This, then, is what the WHO (2000) refers to as 'new universalism':

> *Rather than all possible care for everyone, or only the simplest and most basic care for the poor, this means delivery to all of high-quality essential care, defined mostly by criteria of effectiveness, cost and social acceptability. It implies explicit choice of priorities among interventions, respecting the ethical principle that it may be necessary and efficient to ration services, but that it is inadmissible to exclude whole groups of the population. (p.xiii)*

Table 5.1 provides an overview of the scope of policy conflict in the debate about the global health crisis

Structure of policy conflict in the health debate

We have seen how, once again, the three policy stories lay out a space in which policy debate about the health care provision takes place. Since each story frames the health policy issue in a particular way, conflict is likely to be the order of the day. However, this three-way debate also creates rugged patterns of agreement and disagreement.

Areas of agreement: broad principles, common policy measures and mutual rejection

How do the patterns of agreement and mutual rejection observed in the previous chapter play out for the debate about the global health crisis? Both the choice and the stewardship stories make out a close relationship between economic growth and health (World Bank, 1993; WHO, 2000; ILO, 2007). Since exploding health care costs threaten economic growth, both advocacy coalitions favour policy measures that help control costs. For example, proponents of both advocacy coalitions would like the public sector in health care provision to 'steer more and row less'. Both agree that the public sector should concentrate on providing through, for example, education, the right conditions for high-quality health services (World Bank, 1993; Anonymous, 2004a). Likewise, advocates of choice and stewardship see an active and larger role for private sector providers in modern health systems (WHO, 2000; BAK, 2003; BMA, 2004; ILO, 2007).

Since wealth begets health (and vice versa), proponents of the two stories take a rather dim view of the primary health care movement's wanton disregard for the economic realities. Demanding radically democratic primary health care facilities is utterly impractical, not to mention enormously costly. Moreover, both coalitions would agree that offering everyone the same, invariably low-quality, health services is counter-productive.

Advocates of the rights and choice policy stories champion patient rights. At present, both advocacy coalitions agree, physicians who dominate health systems treat patients with paternal and not always well-meaning condescension. In reality, both the choice and rights advocates argue, terms such as 'clinical needs' and 'physician autonomy' obscure a massive infringement on our liberty and rights (not to mention our wallet). Consequently, rights and choice advocates will fight for measures that expand patient rights. First, proponents of both policy stories support increased patient involvement in health policy processes. Measures could include establishing means of redress (such as health system watchdogs or ombudsmen) or including patients in policy processes (e.g. placing patient groups in regulatory gremiums). Second, members of both coalitions support measures that strengthen patient rights indirectly by controlling providers. Policies here comprise a wide range of financial controls such as DRGs, quality targets, league tables and the like. Last, both the rights and choices advocates want to empower individuals and citizens to adopt healthy lifestyles thereby obviating dependency on professionalized health systems. Since centralized health systems do little for patient rights, both advocacy coalitions see the decentralization and devolution of health systems as a necessary precondition for expanding patient autonomy.

Proponents of the health rights and choices stories agree that large and centralized health systems run by and for a technocratic elite are to blame for much of the global health crisis. In these unresponsive, inefficient and dehumanizing systems, patients are mere objects to be processed and fleeced.

Proponents of the stewardship and health rights story recognize that health is a social right. Good health, much like literacy, is a basic precondition for citizenship (Marshall, 1950). For this reason, so both advocacy coalitions argue, the public sector is duty-bound to provide citizens with access to adequate health. Since health care provision is primarily a public service, proponents of the rights and stewardship story welcome health policies that strengthen the public aspect of health care provision. This includes classical public health tasks such as vaccinations as well as maternal and child health programmes. Given new health challenges such as HIV/AIDS or ageing, so contenders from both advocacy coalitions argue, health care provision needs to become part of a general and integrated package of citizen services (Walker, 2002; WHO, 2002).

Consequently, both the rights and stewardship discourses reject the privatization and commercialization of health care provision. Privatization invariably increases costs without increasing in access or quality (Schiff et al, 1994). Commercial health care, both rights and stewardship advocates agree, tends to undersupply public health goods (e.g. vaccines) in favour of unnecessary but profitable health services (Botox, teenage cosmetic surgery). Above all, health care providers driven by profit have all the incentives to exploit inevitable information asymmetries between patients and physician (BMA, 2004).

Intractable disagreement

The three policy stories, carved from fundamentally different frames, create an awkward geometry of agreement over broad issues and general policy measures. As in the other debates analysed in this book, agreement on basic issues and policy measures dissolves into intractable controversy about practical details.

Stewardship vs choices

Stewardship and choices advocates disagree deeply about the relative roles of individual responsibility, the private sector and consumers on the one hand, and public interest, the public sector and providers on the other. For proponents of stewardship, individual choice, while clearly important, is always subordinate to public interest considerations. The British Medical Association (BMA) argues that individuals should.

> ... *be involved as partners in their care at all levels of individual interaction. It is right that they have the opportunity to take different pathways through the healthcare system. However, it is important that these decisions are framed by higher order choices that have been agreed at collective levels. (BMA, 2004, p10)*

Rather than being subordinate to expert decisions, so champions of the choices story argue, individual choices should drive health care provision

(Anonymous 2004b; Hansen, 2007). Hansen (2007) argues that the best way of getting health systems back on track is to '... give patients/consumers back responsibility for their own health, and control over their own health care expenditures'.

By the same token, opinions about the relative weight of public and private sector provision diverge considerably. For champions of choices in health, the share of public health provision is far too large. The Economist points out that ...

> *... just because governments finance so much of health care, they do not necessarily have to provide it themselves. Britain's NHS employs 1.4 million people, making it the world's third-biggest employer, surpassed only by China's Red Army and the Indian railways. (Anonymous, 2004c, p17)*

Health systems would improve immensely, proponents of choices contend, if a large part of this public provision were handed over to more efficient private providers. Although steering more and rowing less, as proponents of the stewardship story claim, may imply allowing the private sector to provide certain health services, this by no means obviates the need for public sector provision. On the contrary, the WHO (2000) argues that a '... strong public sector may even be a very good strategy for regulating private provision and for consumer protection, if it helps to keep the private sector more competitive in price and quality of service'.

The individualist and hierarchical frames also lead to a fundamentally conflicting view of health care providers. Ideally, so proponents of the choices story argue, health care professionals should be services providers as in any other competitive industry (Anonymous, 2004b; Hansen, 2007). Here, the needs and wants of rational consumers should drive provision. On this view, professional autonomy is relevant only when it enables health care providers to respond efficiently to consumer demands.

All well and good, counter the champions of stewardship, if health were a market like any other. But, as we have seen, health is different. Information asymmetries and uncertainties mean that, in addition to being producers of health services, health care professionals are also responsible for guiding patients through the health system (BMA, 2004; Blunden and Smith, 2005). The cohort study of the BMA (2004) reveals that 88 per cent of the doctors who took part in the survey believe they 'should act as patient advocates and guide them through the system' (p7). Only 22 per cent of doctors agree that 'patients are consumers of health and doctors should respond to their demands' (p7).

Rights vs choice

When vocalizing their support for patient rights, the members of each advocacy coalition have very different things in mind.

For proponents of the choice story, patient rights are simply consumer rights. Empowering patients qua consumers simply entails giving them control over health care expenditures. For example, getting rid of third-party financing, Hansen (2007) insists,

> ... means that the patient acquires full consumer power since he is put in control of all health expenditures on his behalf. The crucial point of a consumer-driven healthcare system will be the direct link between personal lifestyle and individual control of health expenditure with fully informed consumers. (p5)

By the same token, producers of health services ought be left alone to innovate, compete and meet demand in the most efficient way.

Advocates of the egalitarian story see patient rights as an integral part of citizenship. Exposing people to the financial risks of health care has nothing to do with granting health rights. On the contrary, it undermines health rights. Denying health care or even questioning the unconditional universal access to health care is tantamount to depriving people of citizenship. Since health care, then, is so constitutive of citizenship, it only makes sense to bring it under democratic control (Werner, 2003; Sanders, 2003). Indeed, the more democratic, the better: PHC, the delegates of the Alma-Ata conference surmised, '... requires and promotes maximum community and individual self-reliance and participation in the planning, organization, operation and control of primary health care' (WHO and UNICEF, 1978, p1).

Similarly, the two advocacy coalitions disagree about what is to count as a public health good. Advocates of choice clearly circumscribe public health goods as those goods and services with positive health-related externalities. This includes services that the market does not supply (e.g. vaccinations, sanitary infrastructure, etc.) or services that help imperfect health markets work more efficiently (e.g. consumer information about costs and benefits of health interventions, provider regulation, etc.).

Proponents of the health rights story take a far more expansive view of public health goods. Here, a public health good is any service or good that empowers, liberates and emancipates citizens. This can include very specific health interventions (such as safe and easily accessible abortions), general infrastructural measures (such as barrier-free public transport), general education (that empowers people to live healthier lifestyles), specific educational programmes (such as nutritional education in schools), environmental policy (that protects healthy natural environments) or social policy (that ensures that we have healthy social environments).

Advocates of the rights and choices discourses have very different ideas of what a healthy lifestyle entails. For the individualists, healthy lifestyles are all about individual restraint and smart choices. This includes moderation in consumption, regular exercise, the intelligent use of medicine (including smart drugs) as well as stiff competition to sharpen the mental faculties. For the proponents of the rights discourse, a healthy lifestyle involves far more than a subscription for the gym and prescription for Prozac. Since an unjust, oppressive and environmentally rapacious world order is making us sick, this is what will have to change if we are ever to be healthy. Thus, we must end all inequality and start living in harmony with our real social, economic and environmental needs.

Stewardship vs rights

Looking at the global health crisis through the hierarchical and egalitarian lenses lets policy actors understand that health is a right. But what that exactly means and, more importantly, what implications this has for policy, is hotly contested.

Proponents of the health rights story see health as an absolute, inalienable and indivisible human right (WHO and UNICEF, 1978; Sanders, 2003). Since real human well-being depends on the health of our social, political and natural environments, real human health implies healing the many wounds caused by an unjust and rapacious capitalist system. Real health policy, then, is '... a part of the long and very formidable struggle to have a just world order' (Banerji, 2003, p13).

A just world order, proponents of the stewardship story agree, is a laudable and worthy cause. But, the proponents of the stewardship story object, given human history so far, it may not be available today. Or tomorrow. Or even next week. In the meantime, there are broken limbs to be set, babies to be delivered, vaccines to be administered, cavities to be filled, chemotherapy to be supervised, diarrhoea to be treated, check-ups to be done ... the list goes on and on. In the light of these tasks, so proponents of stewardship argue, it makes little sense to turn health care provision into an ambitious (and probably unnecessary, but definitely undoable) programme of social, political and cultural emancipation.

In fact, argue proponents of stewardship, this is the reason many of the PHC experiments failed. Believing that health wants are artificially imposed by an oppressive system, PHC systems concentrated solely on the provision of putative 'real health needs'. As a result, the WHO (2000) argues, PHC systems '... foundered when these two concepts did not match, because then the supply of services offered could not possibly align with both' (p.xii).

Much of this egalitarian holism, argue advocates of stewardship, is founded on fundamental misconceptions. First, health rights cannot determine, but merely inform, health care provision. Health care provision, they maintain, is about getting health services to people when they need them and where they need them. This requires the balancing and optimizing of a wide range of factors for any given structural, economic and political context. In this process, Paul Hunt (2004) tells us, the '... indispensable contribution of human rights is that they help to ensure that the interests of the relatively powerless are not neglected but given due weight throughout the balancing process' (p70). Second, just because the ultimate determinants of health are social, political and cultural, it does not follow for hierarchical actors that health care provision must reconstruct human civilization. Given the already formidable tasks at hand, it makes far more sense for health policy to concentrate on more immediately health-related concerns and issues. Third, balancing the process also means coming to terms with socio-economic and political realities as they are, at least in the short and medium run. On this pragmatic view, strategic alliances and partnerships with a wide range of different policy actors are necessary in order to muster sufficient resources for health care provision. That is why public–private partnerships, for example to provide access to medicines, may not be ideal but they are nonetheless essential.

For proponents of the health rights story, contact causes contamination. The qualifications and rationalizations that we keep hearing from organizations formerly committed to a PHC agenda are merely excuses. They prove, so the argument goes, that the global forces of oppression have usurped and corrupted these institutions (John, 2003; Banerji, 2003; Sanders, 2003). In the past decade or so, both UNICEF and the WHO have sold out to the 'global forces behind the dominant inequitable paradigm of development' (Werner, 2003, p18). Have these 'public-private

Table 5.2 *Structure of policy conflict about the global health crisis*

	Choices – Stewardship	Rights – Choices	Stewardship – Rights
Agreement	Health implies wealth; Costs need to be controlled	Patient rights; Measures to strengthen patient rights	Health is a social right
Mutual rejection	Sever link between wealth and health	Centralized and hierarchical systems of health provision	Commercialization and privatization of health
Disagreement	How to go about securing prosperity	Precise nature of patient rights	Granting social rights in practice

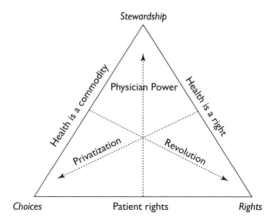

Figure 5.2 *Triangular policy space of the global health crisis debate*

partnerships' and 'joint projects', ask proponents of the health rights story, brought any relief to the unjust disease burden in the developing world? The answer is a resounding: No. 'To the contrary', argues Werner (2003), 'these new partnerships by UNICEF and WHO with transnational corporations further entrenches [sic] and legitimizes the forces that put healthy profits before people' (p17).

Table 5.2 provides an overview of the areas of agreement and disagreement between contending advocacy coalitions. Again, the scope and structure of policy conflict about the global health crisis generate a triangular policy space that defines and delimits the debate (Figure 5.2)

Potential impacts of conflict in the health care debate

As we have seen, the frames impose order onto the complexities of health care provision by selectively emphasizing, stressing and foregrounding some aspects while pushing other dimensions and factors into the background. It is precisely this feature of frames that allows policy actors to tell coherent, plausible and convincing policy stories about health care provision. By focusing and foregrounding certain factors of health care provision, however, each policy story comes with in-built blind spots. Each frame sharpens policy-makers' awareness and understanding of specific relationships in health care provision at the price of obscuring from analytical view other central causal chains. As a result, frames and policy stories are powerful cognitive and explanatory tools that, however, come unstuck when faced with problems that emerge from within the frame's blind spots.

Choices and inequality

The choices story tells a tale of how health thrives in competitive environ-ments. In telling their tale, advocates place an inordinate amount of trust in the workings of the market. While proponents of the choices story are acutely aware of inequities created by public health systems that tamper with the invisible hand, they are less aware of the inequality generated by markets.

First, they consistently downplay the level of resources needed to be a rational market actor in health care. For example, using the web as a source of information on cost-effective treatment, as Hansen (2007) or the World Bank (1993) suggest, presupposes that patients not only have access to the web but, more importantly, they also have the skills to extract the right kind of information at the right time. In developing countries, basic literacy, let alone digital literacy, is an additional formidable barrier to rational market participation.

Second, and this is something advocates of choice acknowledge (World Bank, 1993), the well-known imperfections that afflict markets for health care require sophisticated administrative skills and resources. Introducing private sector health care provision replaces public-service orientations, however corrupted, with the profit motive. This means that health care provision will follow the money. And if this money is not adequately chan-nelled and disciplined by functioning regulation, it is likely to concentrate on providing unnecessary but profitable health services and undersupply public health goods. In a very real sense, then, strong markets for health care require strong states to regulate them.

Third, proponents of the choices story also overestimate the role of indi-vidual responsibility in health and disease. The underlying assumption is that, with very few exceptions, individuals are responsible for their health and that disease results from specific patterns of behaviour. This ignores the very many causes of disease that are unpredictable, partly because of the inherent complexities of human health (for example, the interaction between genetic predisposition, the environment and individual behaviour) and partly because many causes of diseases remain unknown.

Fourth, the fundamental premise of the choices story is that the mar-ket regulates the provision of health care through the price mechanism. In practice, this means that, in order to prevent waste and inefficiency, the price for some health services needs to rise to encourage rational choices in health care provision. Quite simply, the primary impact of policy pro-posals of the choices story is that health services will, on the whole, become more expensive. But people cannot always choose (or predict) what disease they will suffer from or when they will be felled by an accident. What is more, some people, such as children or the disabled, require considerably more medical attention than healthy adults. Market reforms, then, unfairly expose these people to financial risk.

Last, while not completely unaware of these potential impacts, proponents of the choices story consistently underestimate the proportion of vulnerable and disadvantaged groups and overestimate the size of the 'self-financed' health system. Hansen (2007) mentions, en passant, that the state will need to finance health services for about 10 per cent of the population, the deserving poor, that genuinely cannot afford health care in developed countries.

Rights and coercion

Underestimating the size of the disadvantaged and downtrodden is a mistake unlikely to occur to proponents of the health rights story. However, just as the advocates of choices unduly emphasize individual behaviour as the cause of disease, proponents of the health rights story reduce the complexities of health and disease to socio-cultural factors.

As a consequence, proponents of the health rights story downplay the role of medical expertise on the one hand and self-interest on the other. While many diseases originate in a person's environment, many others are of genetic origin. To complicate matters further, some diseases are partly caused by genetic predisposition and partly caused by environmental factors. The great unknown, as we have seen, is what causes what, under what circumstances and to what degree. In addition, serious diseases require more than basic primary health care administered by semi-professional or traditional healers. Proponents of the health rights story conveniently forget that it is precisely the inequitable economic order and physician-dominated techno-medicine that enables most children in the developed world to see their fifth birthday. By the same token, advocates of health rights completely discount the importance of wants in health care. And yet, distasteful as it may be, it is the profit motive driven by health wants that provides at least some of the impetus for developing effective drugs and therapies.

While good at seeing the 'big picture' of how our social, political and natural environments impinge on our health, advocates of health rights tend to be less interested in the minutiae of health policy and health care provision itself. While there is much truth in the argument that contemporary health provision and health policy treats the symptoms rather than the underlying causes, this does not mean that we should not treat the symptoms as and when they appear. Given the immediacy and sheer dimensions of the global health crisis, the holistic orientation of the health rights story seems of peripheral relevance to most acute health policy issues at hand. Knowing that global socio-economic and political inequities are the true cause of a child's dysentery is informative, enlightening and perhaps even emancipative. It will, however, do nothing to stop the child dying of dehydration within the next week.

Similarly, the exacting moral standards that advocates of health rights apply to health policy measures cloud their appreciation of incremental but highly effective health interventions. On this view, health care is legitimate if it emancipates and empowers the poor and vulnerable by undermining the grossly inequitable socio-economic world order. This, however, disqualifies a very long list of health interventions that provide short-term health benefits without, however, bringing about a better world by tea-time.

As a result, proponents of the health rights story are vulnerable to a number of weaknesses. First, their insistence on moral purity can lead to radicalism. By blaming disease on the corrupt socio-economic order and its ruling elites, proponents erect a 'wall of virtue' around the like-minded. At the same time, it depicts anyone who may disagree with the health rights discourse as, at best, morally compromised, or, at worst, irredeemably evil. Here, as we have seen, good health is synonymous with salvation from the oppressive socio-economic world order. Health policy becomes an emancipatory struggle against the 'forces of oppression', replete with martial rhetoric of armed resistance. Since any contact with the forces of oppression implies contamination and pollution, advocates of health rights tend not to understand the potential of strategic and pragmatic alliances. Given the complexity and resource-intensity of health care provision, this is a considerable disadvantage.

Second, policy measures based on health rights alone are vulnerable to free-riding. By placing excessive trust in people's potential for civic virtue, proponents of the health rights story tend to discount the need for preventing abuse of universal primary health care provision. More importantly, however, proponents of health rights consistently underestimate the degree of resentment that free-riding generates among the target population. Rather than dispelling racism and xenophobia, perceived free-riders, more often than not foreigners or members of an out-group, often become victims of precisely those acts of aggression that universal access to health care was supposed to do away with.

Stewardship and misappropriation

As we have seen, proponents of the stewardship story believe that health care provision is about striking balances. Unlike proponents of the other two health policy discourses, the stewardship story sees no problems in concentrating considerable power in the hands of a few health policy stewards. This confidence in the technical and moral superiority of health system experts – which really means physicians – blinds advocates of stewardship to a number of risks with centralizing power in the hands of physicians.

First, similar to proponents of the health rights, hierarchical actors tend to underestimate the motivating force of self-interest in health care provision. Health care providers, particularly physicians, are not immune to the lures of financial gain and status. On the contrary, in many European countries, the number of GPs is in constant decline as young doctors look to more lucrative and prestigious specializations (BAK, 2008). The brain drain of health professionals from poor developing countries to more lucrative health markets in the developed world does not exactly attest to the public interest orientation demanded by the WHO (Mullan, 2006). In short, health system stewards in many countries may lack the basic qualities of stewardship.

Second, the institutional architectures that enable health care stewards to balance the different contingencies of health policy-making also introduce considerable inertia into health policy-making. Like other institutions of the welfare state, health systems have proven to be very resilient institutions. On the one hand, policy and institutional choices in the past constrain current choices: institutional path-dependencies mean that radical change is both difficult and costly (Bonoli and Palier, 2001). On the other hand, health systems are also political systems that determine who gets what, where and when. In doing so, health systems empower policy actors interested in maintaining the status quo. For well-functioning health systems, this is an admirable characteristic. Unfortunately, institutional resilience and resistance to change also apply to systems in which health governance lacks the qualities of stewardship. In these systems, institutional trust in health care professionals turns into a formidable barrier to necessary reforms.

Third, advocates of stewardship consistently underestimate the contribution of other health care professionals. Traditionally, health care systems have been dominated by physicians both in the surgery and in policy-making. A particularly stark example of physician dominance is the process of health budget negotiation in Japan. The three-stage process of negotiating the so-called unified fee schedule tightly choreographs negotiations between government and the Japanese Medical Association. This negotiation process empowers the JMA and ministry to the detriment of other actors in the health care subsystem (Ikegami, 2005). The problem is that the traditional dominance of physicians belies the increasing importance of other health care professionals in an ageing society. The management of disease, rather than a purely curative focus, requires a range of different competences including physiotherapy, psychotherapy, occupational therapy, pharmacy as well as general skills in the provision of care, not to mention social service competences. Indeed, as proponents of both the choices and health rights story contend, the highly trained and, more importantly, highly remunerated physician may actually play little more than a cameo part in the case management of older patients (Walker, 2002).

Table 5.3 provides an overview of the different weaknesses and vulner-

	Choices	*Rights*	*Stewardship*
Trusts	Markets and competition	Malleability of social norms; inherent good of people	Authority and probity of health professionals
Downplays	Inherent market distortions in health	Self-interest of actors	Dangers of concentrating power
Vulnerable to	Increasing inequality	Free-riding	Bureaucratic inertia and corruption

abilities implicit in contending policy narratives.

Table 5.3 *Potential impacts of policy conflict about the global health crisis*

Conclusion

By applying the typology of frames introduced in Chapter 2, this chapter has modelled the scope, structure and impact of policy conflict about the global health crisis. The chapter has shown that policy actors tell at least three diverging policy stories about health care provision.

The choices story tells a tale of how health thrives in environments where individual responsibility and choice determine health care provision. The story also shows how misallocation and inefficiency lead to suffering where choice and responsibility are absent. The best thing for health in both the developing and the developed world, proponents argue, is to let the market do its magic. This may, proponents readily admit, seem a little cold-hearted at times. But it is only by single-mindedly pursuing their self-interest that health care providers will innovate to produce high-quality health services.

The health rights story describes a world in which pervasive socio-economic and political inequities are making us sick. An unspeakably unjust socio-economic order has closed down access to health care for the vast majority of the world's population. In the name of corporate greed, the political henchmen of global capitalism have dismantled public health systems in the developing and developed worlds. In their stead, they have commercialized health care so that it no longer serves our real health needs. Instead of health policy that looks at our social, political and natural environments, we have been given a commercial system that wants to sell us Botox and Prozac. This system, proponents argue, is beyond salvation: what we need is fundamental social, cultural and, most importantly, political renewal.

As we have seen, proponents of the stewardship story believe that health care provision is about striking balances. Yet, balance in the health care system does not appear from thin air. Rather, identifying, achieving and maintaining balances in the health care systems requires consummate skill, experience and vision: something called stewardship. Not everyone is cut out for the job of health system steward. Indeed, only those with the requisite training and experience can appreciate the many complexities and intricacies of steering complex and sophisticated systems of health care provision. What is more, in order to apply their skills and experience, health system stewards need a high degree of autonomy.

The chapter has also shown how the three policy stories give rise to involved patterns of agreement and disagreement. The choices and stewardship stories both tell of the importance of economic growth for health care and are suspicious of plans to abolish capitalism. However, the precise role of the market in health care provision remains a highly contentious issue. Similarly, proponents of the choices and health rights stories can agree that health systems ought to empower individuals. They are equally unconvinced of the need for large, centralized health systems. Yet, on closer inspection, proponents of the choices story mean consumer rights while the proponents of the health rights story are thinking of extensive social rights. Last, the stewardship story and the health rights story agree that health is a right. They frown on reform proposals calling for extensive privatization of health care provision. However, the advocates of the two coalitions cannot agree on how best to secure this right: egalitarians demand radical decentralization and participation while hierarchists call for strong institutions to protect health rights.

Last, the chapter compared the weaknesses and blind spots of each policy story. The unquestioning trust that proponents place in the efficiency of the market leads them to downplay the adverse impacts on equity that the privatization of health care may bring about. The health rights story, in turn, is overly optimistic about communitarian self-determination which leaves them vulnerable to free-riding. Convinced of the superiority of their moral argument, proponents of health rights are prone to intolerance. Finally, the stewardship story believes that only management skills of health professionals and sturdy governance institutions can solve contemporary health challenges. By concentrating almost exclusively on the management of processes, the stewardship story would leave health systems vulnerable to bureaucratic inertia, ineffectiveness and, consequently, corruption.

All three policy stories tell a plausible tale about the global health crisis. All three stories outline credible if divergent policy solutions. Yet, left to their own devices, each would bring about highly undesirable impacts. Since contending advocacy coalitions frame the global health crisis in fundamentally different ways, conflict is inevitable. While this undoubtedly makes policy processes in the health domain rather unsightly, it enables the critical appraisal of policy arguments. Each of the policy frames is a useful lens through which policy actors can critically refract the policy proposals of contending policy stories.

Rather than obstructing the policy process, this is the way policy stories inform the public debate on global health. In a very real sense, policy stories are the vehicle for transporting scientific knowledge into the policy debate. Policy frames allow actors to make sense of the uncertainties and complexities of health care provision. It is policy frames that enable actors to tell stories by using scientific knowledge to build plausible scenarios about health and health care provision. Stories provide some answers, albeit partial and partisan, to the central questions that science helps us formulate but which it cannot resolve. How does globalization affect health systems? What will demographic ageing mean for health care provision? What are policy-makers to do about old and new infectious diseases? How can we address local and global health disparities? What are the advantages and disadvantages of privatizing health care? What are we to do about increasing health care costs? The divergent answers to these questions – embedded in contending stories – and the policy conflict this gives rise to create the concepts and strategies that form the basis for health care policy.

In particular, this means the following:

• As with transport policy and pension reform, conflict in debates about the global health crisis is endemic. Different groups of policy actors use divergent frames to construct partially contradictory accounts of the messy policy challenge of health. This conflict, as unsightly as it may seem, is key to making sense of the global health crisis. Contending policy stories, and the tension they create, are the way in which we as societies can publicly reflect on the global health crisis. Competing frames allow us to decipher and make sense of the vast torrents of knowledge – scientific or otherwise – that flow through the health policy domain. What is more, the wide scope of policy conflict generates a large pool of ideas and concepts about how best to address the global health crisis.

*   Contending policy stories carve out complicated patterns of tenuous pair-wise alliances between advocacy coalitions. As in the other policy domains discussed in this book, agreement is not a load-bearing structure. While contending advocacy coalitions can agree about general principles, this agreement tends to break down when contending policy actors discuss implementation.
*   The policy proposals associated with each policy story potentially lead to 'unanticipated' adverse consequences. The choices story could lead to the deepening of pronounced health inequities. The health rights story may compromise individual civil liberties. And the stewardship story may create unwieldy and unaccountable bureaucratic behemoths. These consequences are unanticipated for members of the particular advocacy coalition. This suggests that avoiding these pitfalls means involving all advocacy coalitions (at least these three) in policy debates about the global health crisis.

References

Angell, C. (2004) 'Excess in the Pharmaceutical Industry', *Canadian Medical Association Journal*, vol 171, no 12, pp1451–1453.

Anonymous (2004a) 'Survey: The Health of Nations', *The Economist*, vol 372, no 8384, p3

Anonymous (2004b) 'Survey: No Reverse Gear', *The Economist*, vol 372, no 8384, p5

Anonymous (2004c) 'Survey: Keep Taking the Medicine', *The Economist*, vol 372, no 8384, p17

Anonymous (2004d) 'Survey: Treating the Symptoms', *The Economist*, vol 372, no 8384, p14

BAK (2003) 'Gesundheitspolitische Grundsätze der deutschen Ärzteschaft', www.bundesaerztekammer.de/page.asp?his=0.5, accessed 23 February 2008

BAK (2004) 'Kriterien der duetschen Ärzteschaft zur Finanzierung einer patienten-gerechten Gesundheitsversorgung', www.bundesaerztekammer.de/page.asp?his=0.5, accessed 23 February 2008

BAK (2008) 'Gesundheitspolitische Grundsätze der Ärzteschaft – Ulmer Papier', www.bundesaerztekammer.de/page.asp?his=0.5, accessed 1 October 2008

Banerji, D. (2003), 'Reflections on Twenty-Fifth Anniversary of the Alma-Ata Declaration', in R. Narayan and P. V. Unnikrishnn (eds) *Health for All Now – Revive Alma-Ata*, Books for Change, Bangalore

Benson, J. (2001), 'The Impact of Privatisation on Access in Tanzania', *Social Science & Medicine*, vol 52, no 12, pp1903–1915.

Blank R. H. and Burau, V. (2004), *Comparative Health Policy*, Palgrave, Basingstoke

Blunden, F. and Smith, T. (2005) 'What Does Choice Mean? How Can it Be Made

More Meaningful?', *Health Policy Review*, vol 1, no 1, pp16–35

BMA (2004) 'British Medical Association written submission to the Public Administration Select Committee Inquiry – Choice and voice in public services', www.bma.org.uk/ap.nsf/Content/choiceandvoice~defining, accessed 1 October 2008

Bonoli, G. and Palier, P. (2001) 'How Do Welfare States Change? Institutions and their Impact on the Politics of Welfare State Reform in Western Europe', *European Review*, vol 8, no 3, pp333–352

Bündnis 90/Die Grünen (2002) 'The Future is Green: Party Program and Principles', www.gruene.de/cms/files/dokbin/145/145643.party_program_and_principles.pdf, accessed 1 October 2008

Epstein H. (2007), *The Invisible Cure: Africa, the West and the Fight Against AIDS*, Viking/Penguin Books, New York, NY

European Commission (2007) *Communication from the Commission to the Council, the European Parliament, the Economic and Social Committee and the Committee of the Regions on the Health Strategy of the European Community*, European Commission, Brussels

EFPIA (2007) *Medicines in Developing Countries*, EFPIA, Brussels.

Freeman, R. (1998) 'Competition in Context: the Politics of Health Care Reform in Europe', *International Journal for Quality in Health Care*, vol 10, no 5, pp395–401

Gilson, L. (1995) 'Management and Health Care Reform in Sub-Saharan Africa', *Social Science & Medicine*, vol 40, no 5, pp695–710.

GSK (2006) 'A Human Race: Corporate Responsibility Report 2006, Corporate Responsibility Report', www.gsk.com/responsibility/index.htm, accessed 2 October 2008.

Gottret, P. and Schieber, G. (2006) *A Practitioner's Guide: Health Care Financing Revisited*, The World Bank, Washington DC

Hansen, F. (2007) 'SelfCare – Essentials of 21st Century Health Care Reform', www.adamsmith.org/think-piece/health/selfcare-%11-essentials-of-21st-century-health-care-reform-2007111640/, accessed 2 October 2008.

Hunt, P. (2004) 'The Right to Health: What Do I Expect from a Pharmaceutical Corporation?' in K. Leisinger (ed) *The Right to Health: A Duty for Whom? International Symposium Report 2004*, Novartis Foundation for Sustainable Development, Basel, pp68–73

Ikegami, N. (2005) 'Japan', in R. Gauld (ed) *Comparative Health Policy in the Asia-Pacific*, Open University Press, Maidenhead

IAPO (2005) 'Policy Statement – Patient Involvement', www.patientsorganizations.org/showarticle.pl?id=810;n=109, accessed 2 October 2008

IFPMA (2007) *The Pharmaceutical Innovation Platform: Meeting Essential Global Health Needs*, IFPMA, Geneva.

ILO (2007) *Social Health Protection: an ILO Strategy towards Universal Access to Health Care*, International Labour Office, Geneva

ISSA (2007) *Investing in People's Health: Towards Strengthening Health Promotion and Prevention in Social Health*, International Social Security Associations – Technical Commission of Medical Care and Sickness Insurance, Geneva

Islam A. and Tahir, M. Z. (1999) 'Health Sector Reform in South Asia: New Challenges and Constraints', *Health Policy*, vol 60, pp151–169

John, P. C. (2003) 'Whatever Happened to Alma-Ata?', in R. Narayan and P. V. Unnikrishnn (eds) *Health for All Now – Revive Alma-Ata*, Books for Change, Bangalore

Marshall, T. C. (1950) *Citizenship and Social Class and other Essays*, Cambridge University Press, Cambridge

Meek, J. (2003) 'Trillion Dollar Disease', *London Review of Books*, vol 25, no 15

Mullan, F. (2006), 'The Metrics of the Physician Brain Drain', *The New England Journal of Medicine*, vol 353, pp1810–1818

OECD (1998) *Maintaining Prosperity in an Ageing Society*, OECD, Paris

OECD (2004) *Towards High-Performing Health Systems*, OECD, Paris

OECD (2006) 'Future Budget Pressures Arising from Spending on Health and Long-term Care', *OECD Economic Outlook*, vol 2006, no 1, pp145–156

OECD (2007) *Health at a Glance: OECD Indicators*, OECD, Paris

Oliver, J. E. (2006) *Fat Politics: The Real Story behind America's Obesity Epidemic*, Oxford University Press, Oxford

Oxfam, VSO and Save the Children (2002) *Beyond Philanthropy: the Pharmaceutical Industry, Corporate Responsibility and the Developing World*, Oxfam, VSO, Save the Children, London

PHM (2000) 'People's Charter for Health', www.phmovement.org/charter/pch-index.html, accessed 2 October 2008

PHM (2004) 'Mumbai Declaration', www.phmovement.org/md/md-english.html, accessed 2 October 2008

Robinson (2008) 'Mythbusting Canadian Healthcare Part II: Debunking the Free Marketeers', www.ourfuture.org/blog-entry/mythbusting-canadian-healthcare-part-ii-debunking-free-marketeers, accessed 11 February 2008.

Saltman, R. B. and Ferrousier-Davis, O. (2000) 'The Concept of Stewardship in Health Policy', *Bulletin of the World Health Organisation*, vol 78, no. 6, pp732–739

Sanders, D. (2003) 'Twenty-Five Years of Primary Health Care: Lessons Learned and Proposals for Revitalisation' in R. Narayan and P. V. Unnikrishnn (eds) *Health for All Now – Revive Alma-Ata*, Books for Change, Bangalore

Schiff, Bindman and Brennan (1994) 'A Better-Quality Alternative: Single-Payer National Health System Reform', *JAMA – The Journal of the American Medical Association*, vol 272, September

Sivaraman, S. (2003) 'Fuzzy Words, Sharp Bullets' – Media, Globalisation and Health' in R. Narayan and P. V. Unnikrishnn (eds) *Health for All Now – Revive Alma-Ata*, Books for Change, Bangalore

Tanner, M. (2006) 'No Miracle in Massachusetts: Why Governor Romney's Health Reform Won't Work', *Cato Briefing Papers*, June, no 97, p1–13

Vasella, D. (2004) 'The Right to Health: The Novartis Position' in K. Leisinger (ed) *The Right to Health: A Duty for Whom? International Symposium Report 2004*, Novartis Foundation for Sustainable Development, Basel, CH

Walker, A. (2002) 'A Strategy for Active Ageing', *International Social Security Review*, vol 55, no 1, pp121–139

Werner, D. (2003) 'The Alma-Ata Declaration and the Goal of "Health for All" 25 Years Later – Keeping the Dream Alive' in R. Narayan and P. V. Unnikrishnn (eds) *Health for All Now – Revive Alma-Ata*, Books for Change, Bangalore

WHO (2000) *The World Health Report 2000: Health Systems – Improving Performance*, WHO, Geneva

WHO (2002) *Active Ageing: A Policy Framework*, WHO, Geneva

WHO (2006) *World Health Report 2006: Working together for Health*, WHO, Geneva

WHO (2007) *World Health Statistics 2007*, WHO, Geneva

WHO and UNICEF (1978) *Primary health care: report of the International Conference on Primary Health Care, Alma-Ata*, USSR, 6–12 September, 1978, World Health Organization, Geneva

WTO Secretariat and WHO (2002) *WTO Agreements & Public Health*, WHO, Geneva

World Bank (1993) *World Development Report 1993: Investing in Health*, World Bank, Washington DC

6

Conclusion – Mess, Conflict and Pluralism

What, then, are we to do about messy policy problems?

This final chapter explores the implications of the preceding analysis for policy-making. After briefly reviewing the argument of the book, this conclusion turns to questions of learning and adaptation. How does frame-based conflict shape policy-oriented learning? How can policy subsystems in the throws of frame-based conflict, such as the ones we have explored in the previous chapters, adapt to change? How do scope, structure and impact of policy conflict shape processes of learning and adaptation? What should policy-makers dealing with messy issues try to avoid and what should they aspire to? By addressing these questions, the chapter distils some general conditions for robust policy-oriented learning and adaptation that follow from the analysis in this book. The last part of this conclusion, somewhat speculatively, produces a system for classifying different types of policy subsystems in terms of the way they enable policy actors to confront frame-based conflict.

Scope, structure and impact of frame-based conflict

We now know a lot more about the way policy actors deal with messy policy issues.

The preceding chapters have developed a framework for narrative analysis and applied it to four messy policy problems. The framework relies on a number of approaches in the social sciences – not least cultural theory inspired by Mary Douglas. It conceives of policy-making about messy issues as an inherently conflictual and argumentative activity. Because of the complexity and uncertainty of messy issues, data and evidence provide little unambiguous guidance to policy-making. That is why policy actors use frames – coherent sets of ideas and beliefs – to make sense of messy issues. Frames allow policy actors to recognize 'salient' facts and provide interpretive templates for weaving these facts into a plausible explanation about what is going on. Individuals that share a particular frame – we have called these groups advocacy coalitions – use frames to tell plausible and convincing stories about messy problems. While these stories are certainly not mere works of fiction, they are selective accounts and specific interpretations of messy issues. Each story, then, is potentially contestable by someone who looks at the same problems through the perceptual lens of a different frame and comes to divergent conclusions.

This book has analysed the frame-based and potentially intractable conflict by comparing competing policy stories. In Chapters 3, 4 and 5, the analyses contrasted contending policy stories in terms of their basic assumptions (the setting of the story), the way they defined the problem (the villains) and the solutions they proposed (the heroes).

Using cultural theory's typology of frames and advocacy coalitions, the preceding chapters dissected intractable policy conflict about climate change, transport, ageing and health as follows:

- Scope of policy conflict: By reconstructing and comparing the policy stories told by the three active social solidarities (individualism, hierarchy and egalitarianism), the analysis has fathomed the argumentative distance between contending policy arguments.
- Structure of policy conflict: The interaction between advocacy coalitions in a dynamic disequilibrium system generates a rugged landscape of agreement and disagreement. Here, the analysis has identified patterns of agreement and mutual rejection across advocacy coalition boundaries. This analysis has also examined the durability of the patterns of agreement.
- Potential impacts of policy narratives: Last, each chapter looked at the potential policy implications of contending policy stories. Since each frame works via selection and interpretation, each policy story foregrounds some aspects of the messy policy story and backgrounds other aspects. The preceding chapters have explored the potential impacts of contending policy stories.

Scope

Chapters 3, 4 and 5 suggest that policy debates about messy issues generate an irreducible plurality of voices and stories. Since each of these stories is based on fundamentally incompatible values, none of them are close substitutes for each other.

The opening chapter introduced the narrative analysis by comparing policy stories about global climate change. For advocates of the profligacy story, unsustainable life-styles in the north are to blame for climate change. Built on horrendous inequities between rich and poor, the exploitative and unsustainable socio-economic system of the north is responsible for almost all of global carbon-dioxide emissions. Saving the planet's climate means ridding the world of this destructive system. For proponents of the prices story, environmental degradation is the consequence of distorted resource prices. Wrong-headed economic policy meddling with market mechanisms, is usually to blame for distorted prices. The rational policy response, then, must be to 'get the prices right'. Not only will this encourage a more efficient use of resources, it will generate much-needed economic growth. Last, the population story tells a tale of how population pressures, predominantly in the south, drive the growth of carbon-dioxide emissions. This story urges environmental policy-makers to adopt a more interventionist approach to managing demographic developments.

In European transport policy, contending advocacy coalitions tell similarly incompatible stories. Advocates of the efficient mobility story believe that the relationship between mobility, opportunity and prosperity is the driving force in transport policy. Once this virtuous cycle spins freely, innovative entrepreneurs will solve problems such as pollution or accidents. Proponents of the sustainable mobility story argue that the inequities and disparities built into modern transport systems are killing us. Hypermobility, they argue, is geared towards serving the greed of the few at the cost of the weak and vulnerable. If transport systems are to serve real human needs for healthy environments, so the argument goes, we need to eliminate unnecessary mobility. This implies reorganizing transport systems from the bottom up. Last, proponents of the balanced mobility story recognize that transport systems are an integral part of our prosperous lifestyles that we can scarcely do without. But, they contend, too much of a good thing brings about all sorts of environmental and social problems. These problems will not go away on their own. And, while radically reducing mobility may work, the cure would be worse than the disease. Instead, careful management of transport systems, they maintain, will give us prosperity as well as healthy and safe environments.

Chapter 4 showed how three contending stories define the debate about ageing and pension reform. The crisis story rings the alarms about pension systems everywhere. Caught in the triple-squeeze of economic globalization, demographic ageing and inherent design flaws, public pension systems are no longer up to the job. Instead of providing adequate incomes for the old while stimulating economic growth among the young, public pension systems distort labour markets, depress national savings and create sizable inequities between the generations. What is needed, proponents of the crisis story argue, are pension systems that allow states to concentrate on the poor by encouraging rich middle-classes to save for their own pensions. Social security, the stability story counters, guarantees social stability and peace. The real problem is that there is far too little social security around the globe. Wherever social security has flourished, as in most of continental Europe, pension systems have been a resounding success. Unfortunately, they continue, this has not stopped the financial industry (through their willing helpers in academia and politics) discrediting public pension systems. In order to advertise their expensive and risky private sector alternatives, these forces have not shied away from systematic deception, disinformation and outright scaremongering. While demographic ageing, economic globalization and social change are real challenges, they argue, competent experts at the helms of social security institutions will, as they always have done, find solutions to maintain stability and social peace. The social citizenship story, in turn, reminds us that demographic ageing will challenge far more than merely pension systems. Ageing, proponents of this story contend, will cause tectonic shifts in almost all areas of our societies: how we will work, how we will live, what diseases we will suffer from and how we will die is set to change profoundly. Such a profound challenge requires a commensurate response. Pension reform can be but one, albeit important element of this comprehensive ageing strategy. What is needed, they argue, is a holistic life-course approach to designing ageing policies. It will require the integration of policy domains as diverse as social services, health care, social security, transport and public infrastructure. More importantly, argue the proponents of social citizenship, it will require ending the discrimination and marginalization of older people.

Chapter 5 looked at the three-way debate about the global health crisis. Here, the choices story tells a tale of how wealth generates health and vice versa. The problem with contemporary public health systems, they argue, is that they do not encourage efficient health choices and corrode individual responsibility. In the developing world, this leads to horribly ineffective and inequitable health care provision as the rich use a disproportionate amount of scarce resource to cure their life-style-related diseases. In the developed world, it leads to sky-rocketing health care costs as people abdicate responsibility for their health to providers and tax-payers. On this view, health care reforms need to empower people by, on the one hand, returning to them the control over their health choices and, on the other, providing real choices in health care provision. For the advocates of the health rights story, today's health systems are, perversely, making us sick. The health rights story shows how an unjust global economic system denies the world's poorest and most vulnerable access to the most basic of health care. At the same time, profit and greed impose unnecessary but lucrative 'cures' onto people in rich countries. Health policy, proponents of health rights demand, needs to stop treating the symptoms and start addressing the underlying social, political and economic determinants of health and well-being. This calls for public participation in health policy at all levels. The stewardship story understands health policy to be a delicate balancing act. Health policy-makers need to square legitimate individual demands on health systems with the provision of basic care to everyone. This requires skill and a sense of proportion that they call 'stewardship'. The problem is that good health policy stewardship is all too rare, particularly in the developing world. As a result, health systems in developing countries are imbalanced, and the health care they dispense is ineffective and inequitable. What is desperately needed here, proponents contend, are institutions and practices that foster stewardship. Astoundingly, health policy-makers are currently undermining systems that have mastered and implemented good stewardship. By introducing one-sided measures for promoting competition, policy-makers are in danger of upsetting the delicate balance in high-performing health systems. That is why policy-makers must immediately stop eroding institutional capacities for striking balances in an extraordinarily complicated policy domain such as health.

The case studies show how policy debate about complex and uncertain problems generates an irreducible plurality of voices and narratives. None of the stories within any of the policy domains can be readily reduced onto another.[1] Neither are the contending stories close substitutes for each other. Each argument is suffused with the kind of interpretation, judgement and values that makes them incompatible, each argument containing facts and evidence that make them hard to dismiss. In each of the domains, the three stories define and delimit a deliberative space – which we can depict in terms of a triangle (see Figure 6.1) – within which policy debate about messy problems takes place. The three narratives or stories also define the pool of ideas from which policy-makers draw when formulating policy responses.

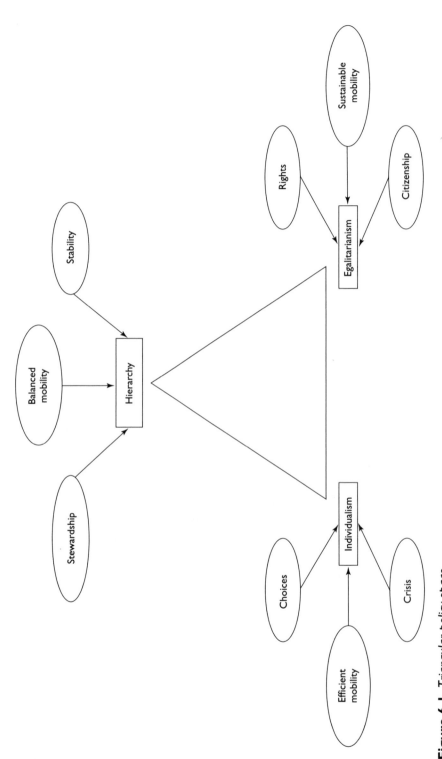

Figure 6.1 *Triangular policy space*

Structure

The structure of policy conflict reveals the volatile nature of interaction between advocacy coalitions. In each policy domain, the patterns of agreement and disagreement make up a complex and inherently unstable array of potential alliances. In all domains, consensus rarely reaches beyond the basic agreement that climate change, transport, ageing and health are issues worth disagreeing about. Yet, in all the policy domains explored in this book, each advocacy coalition found something agreeable in another coalition's narrative. More importantly, pairs of advocacy coalitions agree on the things they definitely reject.

In transport policy, advocates of efficient mobility and balanced mobility agree about the importance of economic growth. They are both suspicious of proposals to radically reduce mobility. The balanced mobility and sustainable mobility stories tell similar tales of the adverse effects of transport on the environment and society. Both are sceptical of policies that deregulate mobility. Proponents of efficient mobility and sustainable mobility agree on maximizing choices for transport users. They both reject central planning and control of transport systems.

When arguing about pension reform, the crisis and stability stories agree that pension systems have an important impact on prosperity. They are less convinced of the need to expand the scope of ageing policy beyond social security institutions. The advocates of the stability and social citizenship stories share a concern for solidarity between the generations. They firmly reject attempts to expose social solidarity to the greed and insecurity of financial markets. The social citizenship and crisis stories both advocate universal, flat rate pension benefits. Both coalitions are highly critical of privileges, say for civil servants or male workers, built into pension systems.

In the conflict about the global health crisis, proponents of the choices and health rights stories champion patient autonomy. This is why both coalitions take a dim view of the way doctors dominate health systems. The health rights and stewardship stories, in turn, agree that health care provision is a public duty. Consequently, they reject the wholesale privatization of health systems. The stewardship and choices stories, in turn, overlap in their concern for meeting legitimate health wants. They dismiss as naive and impractical proposals for doing away with commerce and profit in health care provision.

This three-way contest creates a volatile argumentative environment. In a triangular policy space, alliances between two coalitions are always under threat from two directions. First, agreement fragments 'horizontally' across contending advocacy coalitions. Here, each advocacy coalition shares an area of potential agreement with each rival. Agreement and alliances, then, are fragile since each advocacy coalition can readily 'defect' into another area of agreement and mutual rejection. Additionally, in triangular argumentative spaces, the third coalition is always poised to contest the pair of allies: therefore agreement is always embattled and in need of defence.

Second, since advocacy coalitions enter areas of agreement from different premises, this agreement is brittle. As we have seen, when pushed on details, agreement tends to disintegrate into intractable conflict. For example, agreement between advocates of the sustainability and the rational planning stories about the negative side-effects of transport does not include a consensus about what should be done. Similarly, while champions of the crisis and the stability stories agree that pension systems have a considerable impact on economic growth, they differ about the direction of this impact. Part of solving the global health crisis, proponents of both the health rights and choices narrative contend, is to expand patient rights. However, while the choices story refers to a narrow set of consumer rights, the health rights story demands a very broad array of social rights. For this reason, the basis for embattled alliances does not withstand much probing and reflection. The upshot is that, in debates about messy issues, agreement is possible but inherently unstable.

Impact

The book also compared the potential impacts of each policy story. Since stories are selective interpretations of what is going on, they foreground some things and background others. As a result, each story develops characteristic strengths and weaknesses.

Individualists in the transport debate trust the innovation to overcome any adverse effects of modern transport systems. Proponents of the sustainable mobility story are overly confident of the inherent moral superiority of their argument; all too quickly they dismiss contending stories as morally compromised. The balanced mobility story, in turn, is convinced that judicious management by experts is all it takes to bring about prosperity and healthy environments.

In the pension reform debate, the crisis story downplays the risks of trusting retirement savings to global financial markets. The stability story, in turn, never for a minute doubts the ability of social insurance experts to manage their way out of the triple squeeze. The social citizenship story will have us believe that a basic income for all will not be a disincentive to work.

Advocates of health choices cannot see that private provision invariably increases costs and weakens public health systems. This tends to leave the poorest most exposed to financial risks of disease. The health rights proponents, in turn, overestimate the average patient's tolerance for sustained socio-cultural transformation when all they want is to see a family doctor. The stewardship story unjustifiably associates skills and expertise with moral rectitude, probity and unerring judgement.

All this leaves us in a bit of a bind.

On the one hand, the evidence in the preceding chapters suggests that, in a triangular policy space, debate about messy policy problems inevitably gives rise to frame-based conflict. The way advocacy coalitions make sense of messy issues generates an irreducible plurality of views articulated by contending policy narratives. While none of the narratives the book has explored has been so wrong as to be legitimately ignored, they have been highly selective, thus causing controversy and contention. Further, the triangular policy space creates a rugged and uneven terrain of agreement and disagreement. Here, the strategic alliances, of the kind possible if these were only disputes about who gets what, are inherently unstable. Instead, alliances are fragile and easily succumb to the inevitable contestation from the third advocacy coalition. For messy policy issues, a wide scope of policy conflict means that the 'dialogue of the deaf' and Reformstau are never far away.

On the other hand, a wide scope of policy conflict also helps avoid surprises and policy failure. Each frame not only is a blueprint for designing persuasive policy argument, but it also provides the tools for critically assessing contending policy narratives. Narrowing the scope of conflict risks that the detrimental impacts go undetected and uncontested. Similarly, frames help advocacy coalitions pre-empt external and internal changes. Limiting the scope of policy conflict also reduces ability to recognize different kinds of problems coming over the policy horizon. And once advocacy coalitions spot these changes and problems, policy subsystems should be capable of switching strategies swiftly. This, in turn, requires a reservoir of ideas sufficiently variegated to match a wide variety of problem. Again, narrowing the scope of policy conflict reduces this pool of concepts and ideas.

Learning and adaptation in a world of messy problems

In a world of messy policy challenges – this much policy scientists agree – policy conflict is not always about, as Harold D Lasswell famously put it, who gets what, where and when. Policy conflict about messy issues is at least as much about making sense of it all. What really causes global disparities in health outcomes? Global capitalism headed by evil pharmaceutical giants and greedy health insurance companies? Or is it an unholy alliance of the rich middle-classes, corrupt officials and ruthless local politicians to waste public funds on themselves rather than helping the poor? Or is a lack of strong institutions and effective regulation? As we have seen, the available evidence is compatible with all three interpretations. So while this debate is obviously about who gets what, it is also about trying to figure out what is going on and what to do next. This, then, involves a process of learning.

Sabatier and Jenkins-Smith (1993) define policy-oriented learning as the '… relatively enduring alterations of thought or behavioural intentions that result from experience and are concerned with the attainment (or revision) of policy objectives' (p19).[2] Learning in policy subsystems can take on two forms. 'Single-loop' learning describes the process in which advocacy coalitions improve the means of implementing their preferred policy. Typically, this type of learning takes place within a particular advocacy coalition. 'Double-loop learning' takes place when members of advocacy coalitions reflect and change on the underlying theories and values that drive their policy enterprise. This type of learning requires exchanges across advocacy coalition boundaries. The scope, structure and impact of policy conflict about messy issues implies three stylized scenarios or models of policy-oriented learning and adaptation.

The first stylized model, which we can call the Reformstau-and-meltdown learning, features a wide scope of policy conflict. Here, all three active social solidarities forge policy arguments and work them into coherent narratives about what to do next. Yet, while there is a rich and variegated reservoir of ideas, advocacy coalitions are not listening to one another. Conflict between the contenders is bitter and relentless with no one feeling able to give ground on any issue. Here, frame-based conflict is a 'dialogue of the deaf' in which parties to this dialogue develop '… a remarkable ability, when embroiled in a controversy, to dismiss the evidence adduced by … antagonists' (Schön and Rein, 1994, p5). As a result, policy-making stagnates as conflict becomes a way of preventing rival advocacy coalitions making any gains.

This, of course, describes the stand-off between contending veto-players that Tsebelis calls 'policy stability'. Nothing moves except the volume of the shouting. Eventually, internal or external pressure on the policy subsystem becomes so great that the entire system melts down. Learning takes place retrospectively after the meltdown. Since the costs of learning are very high, this model of learning and adaptation undermines legitimacy of the processes in policy subsystems.[3]

A prominent example of the reformstau-and-meltdown model is the international policy process about global climate change. Joanna Depledge (2006) argues that learning in the climate regime has ground to a standstill because the climate debate has ossified (see Box 6.1 for signs of ossification). Similarly, Verweij (2006) argues that the Kyoto protocol is crippled. The climate change policy process, he continues, is dominated by hierarchical actors without, however, being able to shut out the two contending advocacy coalitions. Conflict is anything but constructive, learning is virtually absent.

Box 6.1 *Signs of ossification*

Joanna Depledge (2006) argues that the climate change regime has become incapable of policy-oriented learning. It has, she continues, become ossified. The following symptoms describe ossification:

- Political entrenchment: Political alliances and coalitions have remained remarkably stable over time. In that time, they have deepened the ideological trenches between the contending parties.
- Taboo topics: The climate change debate, Depledge argues, is characterized by taboo topics that no one dares to address. These include the future commitment of developing countries or the post-2012 regime.
- Stuck issues: Some issues are being addressed in international negotiations but never seem to get resolved one way or another. These comprise, Depledge argues, the issue about 'bunker fuels' or the status of non-Annex I countries from Central Asia.
- Underlying stagnation: This, Depledge contends, describes '... a general sense that the process is stagnating, that old grievances are recurring rather than being resolved, and that, increasingly, the regime no longer constitutes a useful forum for the productive exchange of views' (p8).

One way of avoiding the Reformstau-and-meltdown option, some commentators argue, is to control deliberation by narrowing the scope of conflict. As we saw in the introduction, the proposals suggest empowering governments viz. veto-players.[4] This option leads to what we can call boom-and-bust adaptation.

For messy policy problems, however, such a regime promotes only single-loop learning. As the dominant advocacy coalition implements its preferred policy solutions, it may learn through trial-and-error to perfect its policy instruments. However, after initial success, the dominant advocacy coalition will, to their surprise, increasingly find their policy solutions failing to produce expected outcomes. The limited reservoir of policy solutions means that advocacy coalitions cure problems by administering more and higher doses of the harmful medicine.

For example, the health choices story contends that the solution to poor health coverage in the USA is to further deregulate the insurance and health care industry (Anonymous, 2004a, 2004b, 2004c, 2004d). While everyone is clamouring for less social security benefits, the stability story proposes to significantly expand social insurance pension systems (Gillion et al, 2000). Despite weighty evidence suggesting that most people want to own a car, advocates of the sustainable mobility story would like to compel people to cycle and walk.

What is more, since there is no incentive to respond to contending voices in the policy network, the dominant advocacy coalition will take no notice of the mounting evidence of impending policy failure. In the extreme, the entire policy edifice will collapse under the pressure of sustained and widespread failure. In the wake of the collapse, a rival advocacy coalition will take the helm. Again, double-loop learning, that is changing underlying values and theories of a particular policy, is precarious, sequential and sedimentary; whatever is left of the old regime that the new rulers deem worth keeping will be adapted and ingested. Adaptation is traumatic and costly. Again, the processes of learning and adaptation undermine the functional legitimacy of the particular policy subsystems.

A dramatic example of the boom-and-bust model is the governance of global financial markets that culminated in the financial crisis of autumn 2008. A coalition of individualist actors committed to free markets has dominated the governance of global financial flows for nearly three decades (Balzli et al, 2008). Hierarchical dissenters, including economists such as Joseph Stiglitz (2002) or Paul Krugman (1994), uneasy with what they perceived as the growing recklessness of global financial investors called for more regulation. Egalitarian coalitions, particularly organizations such as ATTAC or Make Poverty History, pointed to the obscene global disparities created by inequitable global financial flows. But since none of these voices had any traction in policy-making, the dominant coalition did not listen. Mounting evidence of policy failure – the dot.com crash in the early part of the century or the housing market crisis in 2007 – was met with more financial deregulation and easy money (Balzli et al, 2008). The collapse of the investment bank Lehmann Brothers in late September 2008 sparked a severe crisis in the global financial system.[5]

Another way of resolving the bind that messy policy issues leave us in is to control the intensity rather than the scope of conflict (Sabatier and Jenkins-Smith, 1993). Box 6.2 (Learning hypotheses) outlines Sabatier and Jenkins-Smith's conditions that promote learning across advocacy coalition boundaries. All four hypotheses are about keeping a lid on the intensity of interaction between advocacy coalitions by enforcing rules and norms of deliberation. Leaving aside for a moment the question of whether Sabatier and Jenkins-Smith's proposals are effective in civilizing conflict, the two policy scientists' hypotheses suggest another scenario of learning and adaptation, one we can call rough-and-tumble.

Box 6.2 *Sabatier and Jenkins-Smith's hypotheses concerning policy-oriented learning*

The advocacy coalition framework (ACF) provides four hypotheses about how learning across advocacy coalition boundaries takes place:

- Hypothesis 6: Policy-oriented learning across belief systems is most likely when there is an intermediate level of informed conflict between the two coalitions. This requires that:
  - each has the technical resources to engage in such a debate;
  - the conflict is between secondary aspects of one belief system and core elements of the other or, alternatively, between important secondary aspects of the two belief systems.
- Hypothesis 7: Problems for which accepted quantitative data and theory exist are more conducive to policy-oriented learning across belief systems than those in which data and theory are generally qualitative, quite subjective or altogether lacking.
- Hypothesis 8: Problems involving natural systems are more conducive to policy-oriented learning across belief systems than those involving purely social or political systems because, in the former, many of the critical variables are not themselves active strategies and because controlled experimentation is more feasible.
- Hypothesis 9: Policy-oriented learning across belief systems is most likely when there exists a forum that is:
  - prestigious enough to force professionals from different coalitions to participate
  - dominated by professional norms.

(Sabatier and Jenkins-Smith, 1993)

Here, the scope of policy conflict is as wide as possible. This means, using the framework applied in the previous chapters, that advocacy coalitions from all three active social solidarities are puzzling over a messy issue. They apply their respective frames to identify problems and draw up plausible solutions. They fashion narratives and recount them to anyone who is willing to listen. And, because of norms and rules about interaction, people from other advocacy coalitions are actually taking in what is being said.

In this scenario, conflict is rife. People from contending advocacy coalitions pick up ideas from their counterparts, critically scrutinize them, change them and release them back into the pool for more scrutiny. In rough-and-tumble learning, surprise and policy failure are less frequent occurrences (relative to the Reformstau-and-meltdown and boom-and-bust): the norms regulating conflict ensure that advocacy coalitions voice objections in ways that impel competing advocacy coalitions to take criticism seriously. And since the rough-and-tumble model has contending advocacy coalitions in constant critical and constructive exchange, when things do go wrong or when external conditions change, alternative solutions are at hand. Adaptation is reasonably swift and costs are relatively low.

As with the other two scenarios, rough-and-tumble is a stylized model. Yet, we can recognize some of its characteristics in European health systems. Here, a broad, frame-based conflict is under way about how best to face old and new health challenges. However, the norms of the medical professions – predominantly but not only exclusively physicians – determine the mode of interaction between contending advocacy coalitions. To be heard by health professionals, members of advocacy coalitions need to 'speak the right language'. For example, in most European countries, contending advocacy coalitions accept, indeed insist on, evidence-based medicine as the foundation for policy deliberation. Health policy, as compared to, say, climate change, is a sector with high-pace innovation. This is not only true for medical and pharmaceutical research, but also for organizational, financial and governance innovations. European health policy-makers have been highly creative in finding ways to increase competition into health provision (e.g. trust hospitals in the UK and Italy). Similarly, recent reforms in Germany have opened health governance to patient organizations.

The heterogeneity of policy actors characteristic of health sectors has meant that the pool of ideas is large. Moreover, rival advocacy coalitions pick up and adapt ideas from their contenders. Box 5.7 in the previous chapter provides a good example. Disease prevention has long been associated with health rights or health choice activists, cf. the Ottawa Charter for Health Promotion (WHO, 1986). Hierarchical health systems, so the argument goes, are incapable of effective prevention because they corrode individual responsibility or they rely on curative (and oppressive) technomedicine. However, proponents of the stewardship discourse are using the idea of prevention to propose a programme of reform for their institutions.

And yet, no one would accuse health policy-making of being harmonious. Indeed, health care reforms in Europe have caused tremendous policy battles. Reforms to promote efficiency are greeted, as we have seen, by howls of indignation from stewardship and health rights advocates. The way competition gets adapted to work in public health systems provokes protest from the health choices crowd. The direction of developments in the alternative medical care field, particularly at the interface of social and health care provision, raises eyebrows in both the stewardship and health choices camps.

And things do go wrong. Starting with the MRSA superbug (in the UK) and ending with scientific dishonesty in medical research (in Austria in 2007) – with a whole lot in between – plenty goes wrong in European health care systems. But none of this leads to the meltdown or implosion of the policy subsystems we see in the international climate change debate or the global banking system. Despite cries of alarm from choices and health rights advocates, European health systems provide high-quality health services to almost everyone who needs them. And this against the backdrop of rapidly changing technological, socio-economic and ecological conditions.

On this view, then, European health policy may be more robust and adaptive than, say, the climate change regime because of its wide scope of conflict as well as the way in which contenders carry out this conflict. The analysis of the scope, structure and impacts of conflict about messy policy issues suggests two general conditions for bringing about effective learning and adaptation:

1   A wide scope of conflict in policy subsystems. In order to avoid surprise and policy failure, the analysis of this book suggests that frame-based conflict needs to feature what Mike Thompson (1996) calls requisite variety. How wide is that? In the approach used to gauge frame-based conflict in this book, it implies including at least the three active social solidarities.[6] This is because, Mike Thompson (1996) argues,

> ... *each way of organizing ultimately needs the others, because they do something vital for it that it could never do for itself. Indeed, this sort of dependency does not have to be mutual; it is enough if each way does something vital for just one of the others and no one of them is left out. (Thompson, 1996, p16, original emphasis)*

In this way, frame-based conflict would be sufficiently variegated for actors to spot the inherent shortcomings of competing policy narratives.

A wide scope of conflict would also increase the pool of ideas and concepts, the material so to speak, used for policy-oriented learning. John Kingdon (1984) likes to speak of this reservoir in terms of a primeval soup of policy solutions. Enterprising policy actors pick up ideas, recombine them with other ideas and see whether the outcome will fly.

A wide scope of policy conflict implies that policy subsystems are accessible. The more open the access to policy subsystems, then, the more robust are learning and adaptation.

2   Frame-based conflict at a medium level of intensity. This, as Sabatier and Jenkins-Smith rightly recommend, implies regulating the interaction between contending advocacy coalitions. A 'dialogue of the deaf' means that competing advocacy coalitions are telling their stories but are refusing to listen to contending narratives. Avoiding intractable controversies, then, has a lot to do with the norms and rules that regulate the way advocacy coalitions deal with one another. And, since policy processes about messy issues are inherently argumentative, these implicit or explicit norms need to regulate the hurling and dodging of policy arguments within a policy subsystem.[7] These rules need to ensure that all voices are heard and responded to.

Thus, the higher the level of responsiveness in a given policy subsystem, the more likely are robust learning and adaptation.

The challenge for policy-makers is to steer policy subsystem processes clear of the Reformstau-and-meltdown and the boom-and-bust scenarios. Instead, policy-makers may want to head for the rough-and-tumble scenario. In order to pilot systems away from either of the politically unsustainable modes of learning, would-be reformers need to know where the policy subsystem in question is in relation to the three extreme scenarios.

In short, would-be reformers need a map of policy subsystems.

Messy issues, learning and pluralism

Accessibility and responsiveness chart a space containing different kinds of policy subsystems. Of course, an institutional space defined by accessibility and responsiveness is nothing new.[8] Readers schooled in the political sciences will recognize the two dimensions from Robert Dahl's famous diagram of polyarchy.

In *Polyarchy*, Dahl uses two basic variables – inclusion and public contestation – to describe differences in political systems. Inclusion determines the level of pluralism in a system. The more inclusive, the more pluralist the polity is likely to be. Public contestation indicates the level of popular control over policy-making. The degree of influence over policy-making points to the degree of democracy in a polity. The higher the degree of political contestation, the more effective is self-determination and, therefore, the more democratic the polity. On this view, pluralism does not necessarily lead to democracy and vice versa.

Pluralist theories explain how conflict between a wide range of actors shapes policy. That is why they seem a sensible place to start thinking about policy-making and frame-based conflict. However, society and politics have changed beyond recognition since American pluralists first formulated their approach in the 1950s and 1960s. On the one hand, Dahl's concept of polyarchy compares nation states, not policy subsystems. On the other hand, pluralists did not much bother with messy issues. Indeed, many of the messy issues, such as urban poverty or crime, surfaced after they developed their theories.

This means that we have to adapt Dahl's original work a little.

Accessibility, political equality and the scope of policy conflict

How does political equality and accessibility relate to policy subsystems?

A policy subsystem, one could argue, endorses the principle of political equality if it is accessible to all stakeholders and citizens. Assessing the degree of openness and accessibility implies examining to whom policy subsystems grant and deny access. By gauging the scope of policy conflict, as Chapters 3, 4 and 5 have done, we can discover whether policy subsystems feature:

- triangular debates in which we find all three active social solidarities
- bipolar debates involving only two of the active social solidarities, essentially a dialogue
- monocentric debates dominated by a single social solidarity, essentially a monologue.

On this view, discovering evidence of only a single policy story suggests a monocentric and therefore exclusive debate. Finding traces of two contending policy stories suggests a bipolar space that is partially inclusive. And the presence of policy stories representing all three 'active' types of advocacy coalitions would indicate an inclusive policy subsystem consisting of a triangular policy space.

A triangular policy debate indicates the widest possible diffusion of political power. Bipolar debates consisting of only two active social solidarities suggest that political power is less diffuse, yet not sufficiently concentrated for a single social solidarity to exert epistemic sovereignty. Also, in monocentric debates, which are not really debates at all, political power is most concentrated: here, members of a single social solidarity have considerable policy-making autonomy in the policy subsystem.

How would this relate to policy-making? Moving rightwards along the accessibility dimension means that a policy subsystem features policy actors from an increasing number of social solidarities. As we have seen, the more social solidarities are present in a policy subsystem, the larger is the pool of ideas for dealing with messy policy issues. The larger the pool of concepts, in turn, the larger the potential for policy innovation, adaptation and strategy switching. Moreover, as accessibility increases, the risk of surprise and policy failure decreases.

Responsiveness, popular control and clumsy solutions

Pluralism is also about the way policy subsystems regulate interaction between advocacy coalitions. The capacity to influence policy-making depends on the way policy subsystems configure policy conflict.

Within a policy subsystem, the quality of debate determines popular control over policy-making. Policy deliberation that enables all participants in policy subsystems to influence policy-making[9] maximizes popular control and self-determination. For all actors to have a fair chance of shaping policy-making, policy deliberation needs to be responsive: that is, policy debate must enable all voices to be heard and responded to. Otherwise, the debate, as accessible as it may be, will degenerate into a 'dialogue of the deaf'.

Policy outputs offer a means of gauging responsiveness in policy debates. Cultural theorists refer to policy outputs that contain elements of all contending policy stories as 'clumsy solutions' (Verweij et al, 2006).[10] Apart from accessibility to the policy subsystem (the necessary condition), clumsy policy solutions depend on the quality of deliberation within a policy subsystem (the sufficient condition): the more responsive policy deliberation is, the more likely it is that policy actors will end up with clumsy policy solutions. This suggests three types of deliberation:

- assertive deliberation, where policy actors aim to assert their policy story over contending narratives
- strategic deliberation, where policy actors interact to more effectively pursue their divergent policy goals
- reflexive deliberation, where policy actors critically scrutinize and reflect on both the means and the ends of policy-making

The structure of policy conflict helps to measure responsiveness. Assertive deliberation between two advocacy coalitions means that neither side is sensitive to potential areas of agreement and, instead, focuses exclusively on areas of disagreement. Strategic deliberation, in turn, implies that contending advocacy coalitions will talk about general policy measures and avoid discussing fundamental principles. Last, reflexive deliberation involves engagement on basic principles as well as on general policy measures.

Reflexive deliberation maximizes popular control by empowering all policy actors in a subsystem to contribute to common policy responses. Strategic deliberation, in turn, limits popular control by focusing on means rather than ends. Assertive deliberation provides least popular control over policy-making since it does not promote interaction.

In terms of policy-making, increasing responsiveness translates into more effective policy outputs. In this model, as deliberation becomes more responsive, each voice in the policy debate becomes more 'audible', clearer and more sensible to contending policy actors. This reduces the incentive to 'shout', thus preventing the decline of debate into a 'dialogue of the deaf'. The higher responsiveness of deliberation, then, the higher the ability to constructively use policy conflict for dealing with messy policy issues.

Regions of pluralism and learning

Plotting accessibility and self-determination in a graph (such as Figure 6.2) gives us a space with three basic provinces, each consisting of three regions. In this terrain, increasing openness broadens the potential for robust policy-making, and growing responsiveness determines whether policy actors can make use of this potential. As we move through this terrain from the bottom left to the top right, the conceptual framework suggests that policy subsystems become increasingly accessible (meaning pluralist) and increasingly responsive (meaning democratic). Policy subsystems found in regions near the origin – the lowlands – are least democratic and least pluralist. Here, rough-and-tumble learning and adaptation to messy policy issues is least likely. Networks along the negative diagonal – the midlands – implement a limited degree of political equality and popular control. In the midlands, learning is likely to be limited and subject to considerable surprises. Last, regions in the upper right-hand province – the highlands – are most accessible and responsive. In these regions, learning and adaptation is likely to be most robust.

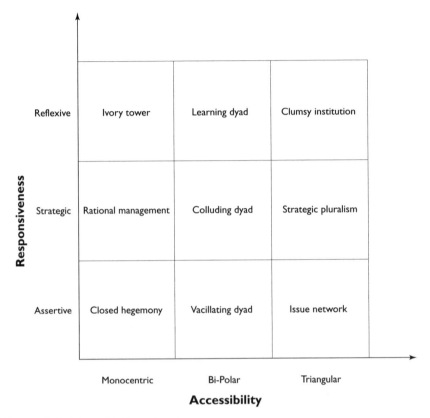

Figure 6.2 *Map of policy subsystems*

The lowlands

In the lowlands, policy subsystems are barely pluralist or democratic. Tightly circumscribed access and unresponsive policy deliberation leave these policy subsystems with a shallow and confined form of democracy.

In the closed hegemony of the lower left-hand corner, a single advocacy coalition dominates: here, might makes right. Policy-makers in rational management networks know what is best for their citizens and will seek expert advice in how to pursue these policy goals. Vacillating dyad, in turn, features two social solidarities locked in fierce and aggressive policy contention.

Lack of contestation and policy conflict mean that decision-making is assertive, authoritative and relatively speedy. However, because effective contestation and deliberation are missing or, if present, are ignored, policy-makers are prone to unpleasant surprises and policy failures. Here, learning follows the boom-and-bust model as policy-makers move from one elegant, but partial solution to another.

In this map of policy subsystems, the lowland region provides few means for learning and adaptation. The lowlands combine the worst of both worlds. In policy subsystems where access is slightly less restrictive, policy debate is fiercely competitive. In networks where deliberation allows single-loop learning, dominance of a single advocacy coalition keeps the pool of potential solutions small. Consequently, policy networks in any of the three lowland regions are static, inflexible and brittle: the small pool of policy strategies and the low level of responsiveness in deliberation offer few options for reacting to socio-economic and political change. In the absence of internal means of strategy switching and adaptation, the model suggests that change, when it comes, is always traumatic and revolutionary. For closed hegemonies, change inevitably means complete collapse, Vacillating dyads generate vacillating policy cycles, and rational management subsystems try to use strategic deliberation to adapt but ineluctably trip up on the narrow scope of policy conflict.

The midlands

The midland region describes policy subsystems that are either more permeable or more responsive than lowland policy communities. In terms of structural features, the midlands encompass somewhat disparate policy subsystems. However disparate, these features give rise to a similar range of effects on policy-making.

The midlands host the two corner-solutions of the top left to bottom right – or negative[11] – diagonal: the issue network and the ivory tower. While the openness of the former provides the largest possible pool of policy stories, the nature of policy deliberation transforms this potential into a cacophony of voices, each trying to outdo the other in terms of volume if not substance. It is here that policy debate is most in danger of degenerating into a 'dialogue of the deaf'. In contrast, the ivory tower possibly features a highly sophisticated and reflexive policy debate without, however, having much to discuss: aspirations to innovation and reform are starved of ideas. The colluding dyad represents the most workable of the midland models since it enables some learning at the strategic level.

Despite differences, the model suggests that midland policy subsystems may all be prone to inertia. Unlike the lowlands, the midlands region supplies policy actors with the institutional wherewithal for learning and adaptation. However, policy networks found in midland regions can make little use of these institutional capabilities. While the ivory tower goes around in deliberative circles, policy-making in the issue network gets lost in the shouting. Unable to thematize fundamental policy goals since areas of agreement across advocacy coalitions are brittle, the colluding dyad collapses into intractability when faced with uncertain policy problems; instead of changing course and adapting policy goals, policy-makers in bipolar cooperative systems apply tried-and-tested policy strategies to new (and essentially unknown) policy problems. This, then, can lead to the Reformstau-and-meltdown scenario of learning and adaptation.

The highlands

The highland region forms a crest along the upper right-hand corner of Figure 6.2. Here, policy subsystems score highest on both political equality and popular control.

Policy actors in the highlands of a pluralist democracy can draw from a deep pool of ideas and concepts. Responsive policy debate allows advocacy coalitions to make the most of this plentiful reservoir. The highland region, then, most effectively avoids surprise and policy failure by enabling swift strategy switching.

Clumsy institutions do best in strategy switching. The triangular policy space ensures that internal slip-ups and developments in the wider policy environment do not go unnoticed. The high degree of responsiveness in the policy debate ensures that these insights (and suggestions on what to do about them) do not go unheard.

In strategic pluralist subsystems, debate also focuses on policy means. Yet, as we have seen in the preceding chapters, triangular debates destabilize and undermine agreement. The inevitable three-way policy conflict injects movement into strategic policy processes thereby reducing the risk of policy deadlock. Given this inherently dynamic constellation, these policy subsystems rapidly reach the limits of strategic deliberation.

The learning dyad also generates a strong current towards clumsy institutions. Reflexive deliberation enables advocacy coalitions to explore all aspects of policy issues. Significantly, this includes the reasons why policymaking keeps getting caught out by unexpected consequences despite wide consultation and deliberation. What is more, deliberation in learning dyads prevents contending policy actors from rejecting out of hand competing and seemingly counterintuitive explanations, solutions and strategies. Policy actors in learning partnerships, we would expect, are more likely than their lowland and midland neighbours to deduce that 'something is missing'. The search for the missing element, then, leads to the extension of the policy subsystem to a clumsy institution.

Policy subsystems in the highlands of pluralist democracy all feature institutional mechanisms and norms for learning and adaptation. While the learning dyad is limited in terms of the scope of policy conflict, strategic pluralism is limited in terms of deliberation. Clumsy institutions, however, overcome both limitations by maximizing the scope of policy conflict. It is in this region that we should expect to find rough-and-tumble learning.

Subsystem democratization and counter-democratization

As with any other map, this map of policy subsystems helps chart journeys across the institutional landscape. Here, movement across the terrain represents change in accessibility and responsiveness of a particular policy subsystem, say, health or pensions policy. For this reason, any institutional change that causes a policy subsystem to 'move' away from the origin potentially increases the pluralist and democratic nature of policy-making. This movement, then, represents a 'democratization' of policy subsystems. There is both an absolute and a relative aspect to this democratization. Absolute democratization describes any journey of change that terminates in the region of clumsy institutions. Relative or partial democratization, in turn, describes developments that shift policy subsystems towards clumsy institutions. Processes of partial democratization can either move a policy subsystem along one of the two dimensions only or can shift policy subsystems incrementally along both dimensions.

However, not all institutional change makes policy subsystems more democratic or pluralist. Any movement towards the bottom left-hand corner of the map depicts a process of counter-democratization.[12] Again, counter-democratization can be both absolute or partial: in the case of the former, institutional change catapults policy subsystems from anywhere in the space to the region of closed hegemonies; in the case of the latter, counter-democratic change moves policy subsystems towards closed hegemony.

Messy policy problems, frame-based conflict and pluralism

This book has tried to make sense of the way policy-makers deal with messy policy challenges. It has done so by taking a close look at frame-based conflict about how best to solve messy issues. The policy sciences have traditionally viewed this conflict with some suspicion. Value-driven conflict, so the argument went, is, as a pathology of the rational policy process, to be avoided at all costs (Stone, 1988). Democracy and pluralism, it is argued, are to be confined to the 'proper' institutional environments so as not to infect rational policy-making with its quarrelsome and corrosive bug of frame-based conflict.

However, even if divorcing democratic politics from rational policy-making were feasible, we have seen that keeping policy subsystems free of conflict is unlikely as long as policy issues are messy. Thus, contemporary policy-making, it would seem, cannot avoid, indeed must embrace, frame-based conflict if it is to tackle the urgent policy challenges of our times. And, unlike many commentators would have us believe, the most effective way of embracing conflict without risking either a melt-down into an intractable cacophony or an implosion due to sustained policy failure is to adopt pluralist and democratic practices in policy subsystems.

Frame-based policy conflict will undoubtedly make policy processes less harmonious or elegant than what we have been used to in the past. But, as Rittel and Webber pointed out in 1973, all the easy and simple problems have been solved; what we are left with is the messy challenges. R. A. W. Rhodes maintains – rightly, I believe – that messy policy problems require messy policy solutions (Rhodes, 1997). The framework developed in this book could contribute to understanding and accepting the way we deal with and adapt to messiness. This, then, may help build untidy (but more resilient) policy processes that yield clumsy (but more robust) solutions to messy policy problems.

Notes

1   Of course, this is not true across policy domains where we see considerable similarities between narratives of the same social solidarity.
2   Sabatier and Jenkins-Smith are much maligned for their 'empiricist' approach to policy-oriented learning (Fischer, 2003). The ACF is highly structured with its parts explicitly laid out. Elsewhere, I have compared their framework to an intricate watch with many moveable parts (Ney, 2006). Of course, since they are explicit about their assumptions and hypotheses, they leave themselves exposed to critical scrutiny.
3   The Reformstau-and-meltdown model resembles what Baumgartner and Jones (1993) have called punctuated equilibria. The two policy scientists argue that US policy change since 1945 has been a story of long spells of policy stability punctuated by large-scale changes.
4   Although this is an extreme conceptual model, this is far more than a straw man. Recall that reforms are under way in Germany, Austria and at EU level to curtail the influence of institutional veto-players, such as the federal states, and political veto-players, such as the trades unions or social partners.
5   At the time of writing, the full extent of the financial crisis is still unclear. Commentators are already declaring the end of an era marked by spectacular growth rate in the global financial sector (Balzli et al, 2008). Some commentators are predicting the end of the global financial system as it has developed in the past three decades. Whatever happens, however, it seems clear that a new financial policy regime will step into the vacuum left by the free market system.
6   Many researchers find the idea of modelling the totality of social life in terms of four-fold typology somehow offensive. Different forms of social and political organization, they argue, are so variegated as to militate being stuck into one of a miserly four categories. What is more, when it comes to frames and ideas, the critics see in cultural theory a restrictive and oppressive programme of categorization and

judgement. 'Telling individuals what to think,' shout the adherents of rational choice. 'Oppressing difference by over-simplification,' shout the postmodernists. This debate, I fear, is an intractable policy controversy of the kind that the preceding chapters have discussed: it will not be resolved in any satisfactory way now or at any time in the future. However, one need not buy into the 'impossibility theorem' to appreciate the fundamental argument about policy conflict and policy learning. What the preceding analysis has shown is that policy conflict about messy issues is by necessity equally messy. Chapters 3, 4 and 5 have demonstrated that public cogitation about messy issues generates at least three incompatible policy narratives. The analysis is agnostic about the possibility of there being more than three irreducible and partially contradictory policy narratives. Indeed, since reality in policy-making is always a lot messier than our models can ever handle, one should expect more incompatible policy narratives than the three that the preceding chapters have outlined. That, however, changes very little about the fundamental conclusions about how policy actors deal with messy policy issues. The upshot is that, while there may be more than three narratives, there probably aren't less.

7   Gianfranco Majone (1989) has thought a lot about this particular issue. Policy deliberation and argument, as it is set up today, seem to aspire to emulate the way scientists converse and argue. Since most of the problems that policy analysts deal with are messy, the scientific model of argumentation, he believes, is not terribly relevant. Worse still, it raises all sorts of expectations about the objectivity of policy arguments that, as we have seen, are difficult to meet in reality. Majone prefers the forensic model: policy processes, he contends, should be organized like legal proceedings.

8   This follows the old political science dictum, attributed to the late Aaron Wildavsky: 'If new, not true. If true, not new' (Thompson et al, 1999).

9   Classical pluralist theory, like much of political science, understands policy deliberation to be about blocking and amending governmental proposals to protect preferences and interests. But, as we have seen, making sense of messy issues also involves constructing narratives that include policy proposals.

10 The opposite, a solution featuring elements from one or two contending narratives, is called an 'elegant' solution.

11 Negative in the sense that it is orthogonal to the one that carries us from closed hegemony to clumsy institution.

12 Interestingly, there does not seem to be an accepted term to denote the opposite of democratization (Pelizzo, 2005), hence the somewhat awkward word.

References

Anonymous (2004a) 'Survey: The Health of Nations', *The Economist*, vol 372, no 8384, p3

Anonymous (2004b) 'Survey: No Reverse Gear', *The Economist*, vol 372, no 8384, p5

Anonymous (2004c) 'Survey: Keep Taking the Medicine', *The Economist*, vol 372, no 8384, p17

Anonymous (2004d) 'Survey: Treating the Symptoms', *The Economist*, vol 372, no 8384, p14

Balzli, B., Brinkbäumer, K., Hornig, F., Hoyng, H., Mahler, A., Neubacher, A., Reuter, W., Pauly, C. and Sauga, M. (2008) 'Der Offenbarungseid' *Der Spiegel*, p20

Baumgartner, F. R. and Jones, B. D. (1993) *Agendas and Instability in American Politics*, University of Chicago Press, Chicago, IL

Depledge, M. (2006) 'The Opposite of Learning: Ossification in the Climate Change Regime', *Global Environmental Politics*, vol 6, no 1, pp1–22

Fischer, F. (2003) *Reframing Public Policy: Discursive Politics and Deliberative Practices*, Oxford University Press, Oxford

Gillion, C., Turner, J., Bailey, C. and Latulippe, D. (2000), *Social Security Pensions: Development and Reform*, International Labour Office, Geneva

Kingdon, J. W. (1984) *Agendas, Alternatives and Public Policies*, Little Brown, Boston, MA

Krugman, P. (1994) *Peddling Prosperity: Economic Sense and Nonsense in the Age of Diminished Expectations*, Norton, New York, NY

Majone, G. (1989) *Evidence, Argument and Persuasion in the Policy Process*, Yale University Press, New Haven, CT

Ney, S. (2006) *Messy Issues, Policy Conflict and the Differentiated Polity: Analysing Contemporary Policy Responses to Complex, Uncertain and Transversal Policy Problems*, PhD Thesis, University of Bergen, Norway

Pelizzo, R. (2005) 'Democracy and Counter-Democracy', Personal Communication, July 2005

Rhodes, R. A. W. (1997) *Understanding Governance: Policy Networks, Governance, Reflexivity and Accountability*, Open University Press, Buckingham

Sabatier, P. and Jenkins-Smith, H. (1993) *Policy Change and Learning: An Advocacy Coalition Approach*, Westview Press, Boulder, CO

Schön, D. and Rein, M. (1994) *Frame Reflection: Towards the Resolution of Intractable Policy Controversies*, Basic Books, New York, NY

Stiglitz, J. E. (2002) *Globalization and its Discontents*, Allen Lane, London

Stone, D. (1988) *Policy Paradox and Political Reason*, Scott Foresman, Glenview, IL

Thompson, M. (1996) 'Inherent Relationality: an Anti-Dualist Approach to Institutions', *LOS-Centre Report 9608*, Bergen

Thompson, M., Grenstad, G. and Selle, P. (1999) *Cultural Theory as Political Science*, Routledge, London

Verweij, M. (2006) 'Is the Kyoto Protocol Merely Irrelevant, or Positively Harmful, for the Efforts to Curb Climate Change' in M. Verweij and M. Thompson (eds) *Clumsy Solutions for a Complex World*, Palgrave, Basingstoke

Verweij, M., Douglas, M., Ellis, R., Engel, C., Hendriks, F., Lohmann, S., Ney, S. and Rayner, S. (2006) 'Clumsy Solutions for a Complex World: The Case of Climate Change', *Public Administration*, vol 84, no 4, pp817–843

WHO (1986) 'Ottawa Charter for Health Promotion', www.who.int/hpr/NPH/docs/ottawa_charter_hp.pdf, accessed 22 October, 2008

Index

Note: A 'b' after the page number indicates the information is in a box; an 'f' indicates a figure; an 'n' indicates a note, the number after showing the number of the note; a 't' indicates a table.